A-Z
PLANT DISEASES

A-Z
PLANT DISEASES

Prof. Gyan Deep Singh

CENTRUM PRESS
NEW DELHI-110002 (INDIA)

CENTRUM PRESS
H.O.: 4360/4, Ansari Road, Daryaganj,
New Delhi-110 002 (India)
Ph.: 23278000, 23261597
B.O.: No. 1015, Ist Main Road, BSK IIIrd Stage
IIIrd Phase, IIIrd Block,
Bangalore - 560 085 (India)
Tel.: 080-41723429
Visit us at: www.centrumpress.com

A-Z Plant Diseases

First Edition, 2009
ISBN 978-93-80106-33-5

PRINTED IN INDIA

Printed at Salasar Imaging Systems, Delhi-110035 (India)

Contents

Preface

Plant Diseases is a complete reference book on plant diseases and their management. Dynamic advances have been made in the last century in plant pathology leading to vast economic benefits. The topics cover biotechnological management of viral diseases, post-harvest diseases of perishables, diseases of seed spices, legumes and mustard. Special chapters on fungi and superficial mycoses, insect and fungal galls and molecular approach for host-pathogen interactions have been included.

Major nematode diseases in plants and their management with reference to India have been discussed. Innovative and novel approaches for the management of plant pests have been described which could be successfully used on a commercial level. India has made rapid strides in the field of agricultural research and education. Green revolution witnessed the use of plant protection umbrella for new varieties and hybrids.

Plant diseases are reasonably uniform in size and stability. Understanding and appreciation of these aspects of the science are essential for the professional plant pathologist, but in a terminal course primarily for agricultural students the practical aspects of the science must receive first consideration. The arrangement of disease within chapters is a compromise in which the systematic relationship of pathogens are given first consideration, but with such modification as is necessary to avoid too abrupt and illogical transitions between disease types.

Author

Chapter 1

Introduction to Plant Disease

The study of plant diseases is known as plant pathology. Infectious diseases are caused by living organisms called pathogens. Noninfectious diseases caused by environmental stress and damage by weather and other environmental factors also will be covered. Indirectly, environmental factors that cause a plant to be stressed may result in the plant's gradual decline. Decline results in the plant being more susceptible to disease organisms.

Because of this, diagnosing plant diseases can be tricky. The real cause of a problem may be the stress factors, with the disease simply being a secondary factor.A plant disease is broadly defined as any condition in which a plant differs in some way from a normal plant in either structure or function. Plant diseases have had an important role in history. The holy fire in the Bible is believed to have been due to the ergot fungus in the heads of grain which made bread poisonous. The great famine in Ireland in the mid-1800s was caused by the destruction of the potato crop by late blight.

WHAT CAUSES PLANT DISEASE?

Plant diseases are generally divided into two groups based on their cause. Nonparasitic diseases are induced by some genetic or environmental factor such as nutrient deficiencies, extreme cold or heat, toxic chemicals (air pollutants, weed killers, or too much fertilizer), mechanical injury, or lack of water. These diseases cannot be transmitted to healthy plants and their control depends solely on correcting the condition causing the disease.

Parasitic diseases are caused by living organisms which derive their food by growing as parasites upon other plants. The most common causes of parasitic diseases are fungi, bacteria, viruses, and nematodes. A few seed-producing plants, such as the mistletoes, can also cause plant diseases.

Fungi are plants that lack the green colouring (chlorophyll) found in seed-producing plants and therefore cannot manufacture their own food. There are between 50,000 and 100,000 different species of fungi of many types and sizes, but not all are harmful.

Most are microscopic in size, but some, such as the mushrooms, are quite large. Fungi reproduce by spores. Bacteria are very small, one-celled plants that reproduce by simple fission. They divide into two equal halves, each of which becomes a fully developed bacterium. Plant diseases are frequently given their common names based on the plant part affected, by the way the plant is affected, or by the cause of the disease.

Viruses are so small they cannot be seen with an ordinary microscope. Many of the viruses that cause plant diseases are transmitted from one plant to another by insects, usually aphids or leafhoppers. Viruses are also very serious problems in plants that are propagated by bulbs, roots, and cuttings because the virus is easily carried along in the propagating material.

Nematodes are small eel-shaped worms that reproduce by eggs. The number of eggs produced by one female nematode and the number of generations in a season depends largely on soil temperature. Therefore, nematodes are usually more of a problem in warmer areas of the country. Most nematodes feed on the roots and lower stem of plants, but a few attack the leaves and flowers.

The life cycles of all disease organisms are greatly influenced by environmental conditions. Temperature and moisture are probably the most important factors that affect the severity of plant diseases. They not only influence the activities of the disease organism, but also affect how easily a plant becomes diseased and the way the disease develops.

How Are Plant Diseases Identified?

The extremely large number of plant diseases (about 50,000 in the United States) makes it impossible for any one person to be familiar with all of them. However, there are certain facts that help in the identification of plant diseases. The most important is to know the name of the plant that is affected. One can then check known plant diseases to see if a similar condition has ever been reported on that plant.

How the disease developed is also important. If it occurred in a very short time (overnight), it is probably not a parasitic disease, but is more likely due to some unfavourable environmental condition or chemical.

How is the condition distributed in the planting? Is it general over the area or only in one or two spots? Parasitic diseases usually do not affect a large percentage of the plants in the early stages, but start in one area and gradually spread to the other plants. And, parasitic diseases usually do not affect several different kinds of plants in one area at one time, even though some disease organisms can attack many different plants.

How Do We Recognize Plant Diseases?

The fact that diseased plants are in some way different from normal plants indicates that there are ways of recognizing plant diseases. Symptoms are outward expressions of plant diseases and include three general types: overdevelopment of tissue, such as galls; underdevelopment of tissue, such as yellowing or dwarfing; and death of tissue, such as blights and cankers.

How Are Plant Diseases Controlled?

The key to controlling parasitic diseases is to introduce an obstacle into the life cycle of the disease organism. This obstacle may be the use of disease-resistant plants on which the disease organism cannot grow. It may be to create environmental conditions that are unfavourable for disease development, or it may involve placing a protective chemical over the surface of the plant.

PLANT PATHOLOGY

DISEASE TRIANGLE

Three critical factors or conditions must exist for disease to occur: a susceptible Host plant, a pathogen, and the right mix of environmental conditions. The relationship of these factors is called the disease triangle.

If only a part of the triangle exists, disease will not occur. Understanding the disease triangle helps us understand why most plants are not affected by the many thousands of diseases that exist.

PATHOGEN

Most pathogens are host-specific to a particular plant species, genus or family. For instance, blackspot of rose will not attack marigolds or lettuce. Some diseases, such as the powdery mildews, produce similar symptoms on different plants. However, the fungi involved are usually host-specific. The rose powdery mildew fungus will not infect zinnias or turfgrass or vice-versa.

SUSCEPTIBLE HOST

A susceptible host has a genetic makeup that permits the development of a particular disease. The genetic defence against a disease is called disease resistance. This resistance can be physical characteristics of the plant (fuzzy or waxy leaf surfaces), chemical characteristics (enzymes that kill pathogens and lack of enzymes) and growth patterns (ability to block off diseased tissue or outgrow damage).

Plants also may be disease-tolerant. Even though infected with a disease, they can grow and produce a good crop or maintain an acceptable appearance. The plant outgrows the disease and symptoms are not apparent or at a damaging level. It is important to remember that plants labeled as disease-resistant are resistant only to a particular disease. They are not resistant to all diseases. Resistance does not mean immunity. Under extreme circumstances, resistant plants may be infected by the disease to which they have resistance.

For disease to occur, the host plant must be at a stage of development that allows it to be susceptible to infection. For example, damping-off only affects seedlings. Botrytis is primarily a disease of buds, although it also can occur on flowers and leaves. Also, it is important that the pathogen be in a proper stage of its development to infect host plants.

ENVIRONMENTAL CONDITIONS

Certain environmental conditions must exist for disease pathogens to cause infection. The specific conditions vary for different pathogens. High moisture and specific temperature ranges, for example, are necessary for many fungal diseases. These conditions must continue for a critical period of time while the pathogen is in contact with the host for infection to occur. Moisture, temperature, wind, sunlight, nutrition and soil quality affect plant growth. If one of these factors is out of balance for the culture of a specific plant, that plant may have a greater tendency to become diseased. For example, lilacs growing in shade are more likely to be infected with powdery mildew than those growing in full sunlight. Environmental conditions also affect the growth and spread of disease pathogens. Very dry or wet weather will have an accompanying set of diseases that thrive under these conditions.

Moisture

Moisture in the plant environment can include humidity, dew, rainfall or water from irrigation. Moisture is critical to the spread of most plant diseases. Familiar diseases, such as black spot, fireblight and apple scab require moisture to spread to and infect new host plants. Constantly wet foliage from overhead watering is a condition that promotes disease development. Seedlings grown indoors in soggy, unsterilized potting medium and pots are more prone to damping-off, a fungal disease.

Temperature

Each disease pathogen has a specific temperature range for growth and activity. There are warm-weather and cool-weather diseases. Many powdery mildew diseases are late summer, warmer temperature diseases. Temperature affects how rapidly pathogens multiply. Soil temperature can also be critical for disease infection. Cool, wet soils promote fungal root diseases. Temperature extremes can cause stress in host plants, increasing susceptibility.

Wind and Sun

The combination of wind and sun affects how quickly plant surfaces dry. Faster drying generally reduces the opportunity for infection. Wind can spread pathogens from one area to another, even many miles. Wind and rain together can be a deadly combination. Windblown rain can spread spores from infected plant tissue, blowing these pathogens to new host plants.

Sunlight is very important to plant health. Plants that do not receive the right amount of sunlight to meet their cultural requirements become stressed. This may make them more susceptible to infection.

Soil and Fertility

Soil type can affect plant growth and also development of some pathogens. Light sandy soil low in organic matter favours growth of many types of nematodes. Damping-off disease increases in heavy, cold, water-logged soils. Soil pH affects pathogen development in some diseases. Clubroot of cabbage occurs in soils with a low pH, for example. High soil pH is a factor in the development of scab on potatoes.

Fertility affects a plant's growth rate and ability to defend against disease. Excessive nitrogen fertilization can increase susceptibility to pathogen attack. It causes formation of SUCCULENT tissue and delays maturity. This can contribute to certain patch diseases in lawns. Nitrogen deficiency results in limited growth and plant stress which may cause greater disease susceptibility.

DISEASE CYCLE

Often gardeners believe that their plants have become

diseased overnight. This may be true in the case of damping-off. More often, however, much has occurred before symptoms are seen. There are five stages in disease development: inoculation, incubation, penetration, infection and symptoms.

The pathogen must be introduced (inoculated) to the host plant. Most pathogens cannot move on their own, but must be carried to the host plant. This is done by rain, wind, insects, birds and people.

Splashing rain carries spores of apple scab fungus from infected apple leaves to uninfected leaves. Wind blows fungal spores from plant to plant. The spotted cucumber beetle transmits bacterial wilt of cucumbers when feeding.

Working in the garden when plants are wet is a common way to spread disease. Disinfesting tools requires a 9-to-1 solution of water and bleach and takes a minimum of ten minutes. Smokers can transmit tobacco mosaic virus from a cigarette to tomato plants.

Seeds or cuttings from infected plants will also transmit disease. Certified seed guarantees that at the time of sale the seeds are free of all diseases. Seeds are often coated with a fungicide to prevent the transmission of surface fungal diseases.Disease-free stock guarantees that the plant is not infected with disease. This is particularly important with perennial plants, such as roses, raspberries and other small fruits.

Incubation

The second stage of disease development is incubation. The pathogen changes or grows into a form that can enter the new host plant. In many fungal diseases, the pathogen arrives on the plant as a spore which must germinate before it can grow into the plant.

Penetration

The third stage is penetration or the point at which the pathogen actually enters the host plant. Once the fungal spore germinates, it sends out thread-like tubes call hyphae. These penetrate the plant through wounds or natural pores.

Wounding roots of bedding plants during transplanting provides entry for root-rotting fungi. The mouthparts of an insect also result in openings for penetration.

Infection

The fourth stage is infection. The pathogen grows within the plant and begins damaging the plant tissue.

Symptoms

As the pathogen consumes nutrients, the plant reacts by showing symptoms. Symptoms are evidence of the pathogens causing damage to the plant. Symptoms include mottling, dwarfing, distortion, discolouration, wilting, and shriveling of any plant part.

PATHOGEN SURVIVAL

Many pathogens can survive without a susceptible host even under the most unfavourable conditions. Many plant diseases survive from one growing season to the next on plant debris, seeds, alternate hosts or in soil.

Because of pathogen survival, it is important to remove and properly dispose of any infected plant materials. It is also important for the gardener to know about the diseases that affect each plant throughout the home landscape, as well as the conditions needed for proper culture.

PLANT PATHOGENS

Pathogens are grouped taxonomically into related groups. It is helpful to know the type of pathogen in order to prescribe a control. Plant disease names are often very colourful and descriptive, as in fireblight, black knot and clubroot. Many disease names indicate something about the appearance of the symptoms.

Fungi

Fungi (plural of fungus) are the largest group of plant pathogens. Fungi often develop into colonies, like bread mold. Some fungi develop into large structures, such as mushrooms. Fungi spread by spores carried by wind, water or animals. They enter host plants directly or through natural openings or wounds. They cause damage by producing substances that change or destroy plant tissues.

Common fungal diseases include powdery mildew, rust, leaf spot, blight, root and crown rots, damping-off, smut, anthracnose, and vascular wilts.

Powdery Mildews

These fungal diseases are very HOST-SPECIFIC. It is common to see powdery mildew on lilacs, roses, zinnias, melons, beans, cucumbers, turfgrass and many other plants. The symptom is a white or gray, powdery growth on leaves and stems.

Powdery mildew will not usually kill a plant; however, it may weaken plants, and it is unsightly. Providing adequate sunlight and air circulation to lower relative humidity and using fungicide sprays for high-profile plants help control powdery mildew.

Rust

Rusts are also host-specific. They produce masses of easily noticed, orange or dark red spore masses on leaf tissue. Rusts are commonly seen in our region on turfgrass, hollyhocks, snapdragons, hawthorns and the small fruits. Rusts usually develop during cool weather.

They are spread by wind and splashing water. Rusts require water to reproduce and infect host plants. Controls include watering early in the day and applying protective fungicidal sprays.

Leaf Spot and Blight

Rose growers are very familiar with black spot, a common leaf-spotting disease of roses. Scab on crabapple and pyracantha are leaf-spotting diseases that can cause defoliation.

Botrytis flower blight occurs on annuals and some perennials during wet seasons. Anthracnose of shade trees is a blight frequently seen on sycamores, maples, ashes and oaks. Fungi-causing tree diseases are spread by wind and splashing water.

Grow resistant cultivars, maintain plant vigor, enhance leaf drying by proper pruning and water early in the day to help avoid leaf spotting and blights. Fungicides provide protection from infection when applied before infection by fungal spores.

Root and Crown Rots

These diseases, which include water mold diseases, attack a wide variety of plants. They cause damage to roots and crowns of plants.

Damage to some of the root system results in poor growth, yellowing or stunting of the plant. Root rot fungal pathogens are found in almost all soils. However, they do not survive as well in well-drained soils. These organisms can live in soil for years in a dormant state. Proper cultural practices, such as correct planting depth and improving drainage, are important controls.

Stem and Twig Cankers

Cankers may be compared to sores on a stem or twig. The tissue may be raised or sunken, discoloured or split open. Twig cankers on junipers and leaf and stem blight of pachysandra are common examples. Most of the canker diseases are not controllable with fungicides.

Maintaining plant vigor, protecting from injury, removing diseased parts and growing plants suitable for the site help prevent canker problems.

Vascular Wilts

Dutch elm disease is the classic wilt disease. The fungus invades the vascular system of the plant and clogs it, stopping the flow of water and nutrients. Verticillium wilt is common on trees, shrubs, vegetables and fruits. Discolouration of the vascular tissue sometimes can be seen when you cut diagonally through a branch. There are no effective chemical controls for wilts. Growing resistant species, maintaining plant vigor and sanitation provide the only controls.

Smuts and Molds

Smut fungi produce large amounts of black sooty spores which show up as unsightly dark brown or black blotches on leaves, stems and fruit.

Bacteria

Bacteria are microscopic, single-celled organisms that multiply rapidly. They produce chemicals that destroy plant tissue or cause abnormal growth. Bacteria enter through wounds, stomata or other natural openings. Bacteria can cause leaf spotting as on English ivy. Fireblight on crabapples, pears, mountain ash, hawthorn, cotoneaster and pyracantha is a bacterial disease. Bacteria also cause diseases, such as crown gall on stone fruits and euonymus and wilts in geraniums and cucurbits (cucumbers, squash and melons).

Disease-free stock and resistant cultivars plus sanitation help prevent bacterial diseases. When cutting out bacterial infections, sterilize pruners with 70% alcohol between cuts. Streptomycin and fixed copper sprays may help slow the spread and development of bacterial diseases. However, chemical controls for bacterial diseases are less reliable than for many fungal diseases.

Mycoplasmas

Mycoplasmas are disease agents that were not accurately identified until the 1960's. Previously many mycoplasma-caused diseases were thought to be caused by viruses. A group of diseases caused by mycoplasmas is called YELLOWS diseases. Yellows disease is usually spread from one host plant to another by an insect vector. Leafhoppers are common vectors for mycoplasmas.

Viruses

Viruses are organisms so small that they can only be seen under an electron microscope, magnified 2,000 to 3,000 times.

Viruses multiply only within living cells of host plants. Viruses are spread by insects, nematodes and humans. Symptoms include vein banding.

Mosaic, flecking or spotting on foliage and abnormal growth, similar to herbicide damage. Tobacco mosaic virus on tomatoes and leaf mosaic on dahlias are common viral diseases. Control is limited to removing and destroying infected plants.

Nematodes

Nematodes are microscopic, worm-like organisms. They most commonly feed on plant roots, but some nematodes invade leaf tissue. Nematodes suck out liquid nutrients and inject damaging materials into plants. They injure plant cells or change normal plant growth processes. Symptoms of nematodes include swelling of stems or roots, irregular branching, deformed leaves, lack of blossoming and galls on roots. Nematodes can facilitate the entry of viruses and fungi into plants. Control efforts are difficult and provide limited success. Chemical controls must be done by professional pesticide applicators. A specialized soil test for nematodes indicates the type of nematode.

DIAGNOSIS OF PLANT DISEASES

To correctly diagnose plant disease problems, follow a few basic steps. View the plant and its environment from various perspectives.

The most obvious place to look first is up close. Use a hand lens if necessary. Don't stop at the first or most obvious symptom; check for more. You want to find all of the symptoms. Look for symptoms on leaves, stems, roots, flowers and fruits. Cut open a branch or stem to look for vascular problems. Vascular problems show as discolouration of vascular tissue, leaf or stem wilting and sudden wilting of a section or a total plant.

General View

Stand back and look at the overall picture. Consider the

total environment: weather, soil, stage of development for plant and pathogens, cultural practices and condition of other plants in the area. A plant growing in the wrong location may be stressed. Consider pesticide applications, recent construction or digging, and weather conditions.

Time

Determine when the symptoms became apparent. The onset of a problem may be due to a cultural practice, the seasonal appearance of a disease or insect, or a weather-related event. Remember that long-term stress is slow to appear, taking a year or more at times. Is the problem spreading? This may indicate it is a pathogen. Are plants of other species affected? Diseases are usually species-specific. Problems caused by environmental factors do not spread, although the symptoms may become more severe.

Knowledge and Experience

You must know what the plant should look like to be able to determine abnormalities. Check the references to see what problems are typical for a particular species. Gather all the information you can to help you make the diagnosis.

Remember that there is usually no single cause. There may be a primary cause; however, it may be associated with cultural or environmental conditions. Just as there is probably no single cause, there is usually no single symptom. Search for all of the symptoms. Orderly thinking and good questions are the key to accurate diagnoses. When in doubt about a diagnosis, turn to agents, state specialists or the Pest and Plant Disease Clinic for assistance or a second opinion.

Nonliving or Abiotic Agents

Nonliving or abiotic agents can indirectly result in plant problems. Additionally, several factors in the plant's environment can produce disease-like symptoms: weather extremes, high winds, high or low temperatures, nutrient deficiencies, physical damage and poor cultural conditions. Frost often damages buds and leaves in early spring. Hail can cause leaf spotting or holes. Drought and high winds result in wilting and in extreme cases, browning and curling. Air, water and soil pollution affect plant health and can produce disease-like symptoms. Soil imbalance, resulting from construction or other dumping, or misapplied garden chemicals can cause damage and disease-like symptoms.

Dumping of household, automotive and industrial chemicals can also produce plant damage. Plant disease can result from a combination of abiotic agents and biotic agents. Plants may be initially placed under stress by nonliving agents. This creates a susceptibility in plants for attack by living agents. Drought may damage roots which then are more likely to be infected by fungal diseases.

Random or Uniform Patterns

Random distribution of symptoms on injured plants is usually caused by a biotic factor, such as infectious disease pathogens or an insect/animal. Uniform patterns are generally associated with abiotic or noninfectious agents like pesticides, fertilizers, environmental or site stress and mechanical damage.

Collecting Specimens

Plant specimens that are to be diagnosed should be taken from the area where symptoms are showing on living tissue. Dead plants are often invaded by secondary pathogens which may hide the original problem. Collect several representative samples showing various stages of disease development. A generous sampling will assist in diagnosis. If possible, collect the entire plant, including roots. Wrap the specimens in dry paper. Do not moisten them or seal them in plastic wrap or plastic bags. Never mix different specimens in a single bag. A fresh sample is required. Complete the diagnostic form as thoroughly as possible. This will result in better diagnosis.

DISEASE CONTROL

By the time disease symptoms appear, disease pathogens are inside the plant and usually beyond control. Therefore, it is important to prevent penetration of pathogens. Consider the following methods of disease control:

Avoid

Avoid certain diseases through choice of appropriate site and planting time, purchase of disease-free stock and cultural practices that do not favour disease infection. Insect control can be important for controlling spread of viruses, mycoplasmas, bacteria and fungi. Mulching can help control diseases by preventing the contact of foliage with soil.

Exclude

Exclude insect vectors with row covers. Don't bring diseased plants into the garden.

Use Resistant Or Tolerant Plants

Appropriate selection of species or cultivars will avoid or minimize problems. Remove all diseased plants, including alternate hosts. Immediate removal of diseased plants or plant parts reduces the chance of the disease spreading. Selective pruning and careful sanitizing of pruning equipment can prevent spread of disease. Rotate crops to avoid soil-borne diseases. Use best cultural practices. Follow recommended cultural practices.

TREAT

Treat plants with recommended chemicals to prevent infection. Some treatments will stop the spread of the pathogen.

Chemical Treatment of Plant Disease

To use fungicides and bactericides effectively, follow these 4 steps:

- Be sure that the diagnosis is correct.
- Choose a recommended control. Remember, a fungicide controls fungal diseases and bactericides only control bacteria. Read the product label and follow the instructions precisely. Make sure the plant you will be treating is listed on the label.

- Apply the control correctly following all label instructions. To be effective, fungicidal sprays must be applied properly. They work by creating a barrier that pathogens cannot penetrate. Good coverage is important. Give special attention to the underside of leaves, especially lower leaves. The barrier's effectiveness is affected by how well the spray spreads and sticks. Many fungicides have SPREADER-STICKERS added to the product. If you are not sure about coverage, observe the pattern spray deposits.
- Apply at the recommended time. In many cases, fungicides are a protectant and must be applied just before the pathogen is able to infect the plant. As the plant grows, additional sprays may be necessary to protect the new growth.

Fungicides are developed so that they degrade fairly rapidly. They are affected by rain, sun, and microbial activity. Excessive rainfall or growth will require shorter intervals between spraying. New shoots should be treated. Product label guidelines usually indicate a range of 7 to 14 days.

Know exactly what disease pathogen or combination of pathogens and abiotic agents are causing a plant problem. Fungicides and bactericides are effective against specific diseases under specific conditions. In the past, general fungicides were used. Many pathogens developed resistance to these general pesticides.

Use common sense in analyzing a situation. If an expensive or sentimental plant is experiencing problems, seek the help of professionals at the Pest and Plant Diagnostic Clinic.

PEST-RESISTANT CROPS

One of the mainstays of integrated pest management is the use of crop varieties that are resistant or tolerant to insect pests and diseases. A resistant variety may be less preferred by the insect pest, adversely affect its normal development and survival, or the plant may tolerate the damage without an economic loss in yield or quality.

Disease-resistant vegetables are widely used, whereas insect-resistant varieties are less common but nonetheless important. Examples include varieties of wheat which have tough stems that prevent development of the Hessian fly and cucurbits (squash, cucumbers, melons) that have lower concentrations of feeding stimulants (cucurbitacins) for cucumber beetles.

Commercial corn lines with increased concentration of a chemical called DIMBOA display resistance to European corn borer, a major pest throughout the world. In the case of cabbage, the only reliable method of controlling onion thrips is through the use of resistant varieties.

Advantages of this tactic include ease of use, compatibility with other integrated pest management tactics, low cost, and cumulative impact on the pest (each subsequent generation of the pest is further reduced) with minimal environmental impact.

The development of resistant or pest-tolerant plant varieties, however, may require considerable time and money, and resistance is not necessarily permanent. Just as insect populations have developed resistance to insecticides, populations of insects have developed that are now able to damage plant varieties that were previously resistant.

CHEMICAL CONTROL

If all other integrated pest management tactics are unable to keep an insect pest population below an economic threshold, then use of an insecticide to control the pest and prevent economic loss is justified. In most cropping systems, insecticides are still the principal means of controlling pests once the economic threshold has been reached.

They can be relatively cheap and are easy to apply, fast-acting, and in most instances can be relied on to control the pest(s). Because insecticides can be formulated as liquids, powders, aerosols, dusts, granules, baits, and slow-release forms, they are very versatile.

Insecticides are classified in several ways, and it is important to be familiar with these classifications so that the choice of an insecticide is based on more than simply how well it controls the pest. When classified by method of application (site of encounter by insect), insecticides are referred to as stomach poisons (those that must be ingested), contact poisons, or fumigants. The most precise method of classifying insecticides is by their active ingredient (toxicant).

According to this method the major classes of insecticides are the organophosphates, chlorinated hydrocarbons, carbamates, and pyrethroids. Others in this classification system include the biologicals (or microbials), botanicals, oils, and fumigants.

Insecticides may be divided into two broad categories: (a) conventional or chemical and (b) biorational. In this guide we define conventional or chemical insecticides as those having a broad spectrum of activity and being more detrimental to natural enemies. In contrast, insecticides that are more selective because they are most effective against insects with certain feeding habits, at certain life stages, or within certain taxonomic groups, are referred to as "biorational" pesticides. These are also known as "least toxic" pesticides.

Because the biorationals are generally less toxic and more selective, they are generally less harmful to natural enemies and the environment. Biorational insecticides include the microbial-based insecticides such as the Bacillus thuringiensis products, chemicals such as pheromones that modify insect behaviour, insect growth regulators, and insecticidal soaps.

The majority of insecticides fall into the chemical category because they are typically more effective and can usually be used to control several pests. This provides the economic justification needed for the research and development of such products. In contrast, the more specialized market of the biorationals makes their long-term economic return less favourable.

Despite the advantages of conventional insecticides, the problems associated with their use have been well documented. These include the resurgence of pest populations after decimation of the natural enemies, development of insecticide-resistant populations, and negative impacts on nontarget organisms within and outside the crop system. One of the more serious problems is the development of resistance. Many insect pest species now possess resistance to some or several types of insecticides, and few chemical control options exist for these pests.

If all other integrated pest management tactics are unable to keep an insect pest population below an economic threshold, then use of an insecticide to control the pest and prevent economic loss is justified. In most cropping systems, insecticides are still the principal means of controlling pests once the economic threshold has been reached. They can be relatively cheap and are easy to apply, fast-acting, and in most instances can be relied on to control the pest(s). Because insecticides can be formulated as liquids, powders, aerosols, dusts, granules, baits, and slow-release forms, they are very versatile.

To compare pesticide use patterns, the dose, formulation or per cent active ingredient of the product, and the frequency of application must also be considered and are used to calculate the EIQ Field Use Rating.

Pesticides with a lower EIQ Field Use Rating have a lower environmental impact. The EIQ concept and techniques are evolving, but this, or a smilar rating, will provide information for appropriate pest management decisions in the future.

Natural enemies are generally more adversely affected by chemical insecticides than the target pest. Because predators and parasitoids must search for their prey, they generally are very mobile and spend a considerable amount of time moving across plant tissue. This increases the likelihood that they will contact the insecticide. When an insecticide is applied, ideally only the target pest(s) should be affected. The goal is to maximize pest mortality while minimizing harm to natural enemies.

The following factors, some of which are used when determining the EIQ, influence the impact of insecticide applications on natural enemies:

- Spectrum of activity. Insecticides toxic to most insects (broad-spectrum) will more adversely affect natural enemies than materials that are narrow-spectrum or selective for specific insect species or life stages. Most insecticides in use today have a broad spectrum of activity.

- Residual (half-life) activity of insecticide. Insecticides that remain toxic to pests for a long time and remain on the treated surface will have a similar effect on natural enemies.
- Coverage and formulation of insecticide. Full coverage sprays will generally have a greater impact on natural enemies than directed sprays, systemics, or bait formulations. Spot or edge treatments directed at localized pest infestations or to a specific plant surface most often occupied by the pest will be less detrimental than those applied to the entire field or plant.
- Dosage and frequency of application. Higher rates and repeated applications will have a more detrimental impact on natural enemies.
- Timing of application. Applying insecticides when natural enemies are not abundant or are less susceptible, such as when immatures are encased in host eggs, can be helpful.
- Susceptibility of natural enemy to insecticide. Some natural enemies are inherently more resistant to insecticides than others, and some populations of natural enemies have been selected either naturally or through the efforts of researchers to possess higher levels of resistance. For example, strains of predaceous mites, rove beetles, lacewings, and species of parasitoids have been selected for increased levels of resistance to insecticides, thereby increasing the likelihood that some will survive insecticide treatments directed at pests.

Some of these pesticide-tolerant strains are commercially available. Applications of some fungicides directed at controlling plant diseases can also reduce the incidence of fungal diseases of insects. In general, little can be done about this incompatibility except to use the fungicide only as needed.

In addition to killing natural enemies directly, insecticides may also have sublethal effects on insect behaviour, reproductive capabilities, egg hatch, rate of development, feeding rate, and life span. Fungicides and herbicides may also have lethal and sublethal effects on natural enemies.

CULTURAL CONTROL

There are many agricultural practices that make the environment less favourable to insect pests. Examples include cultivation of alternate hosts (e.g., weeds), crop rotation, selection of planting sites, trap crops, and adjusting the timing of planting or harvest. Crop rotation, for example, is highly recommended for management of Colorado potato beetle. The beetles overwinter in or near potato fields and they require potato or related plants for food when they emerge in the spring.

With cool temperatures and no suitable food, the beetles will only crawl and be unable to fly. Planting potatoes well away from the previous year's crop prevents access to needed food and the beetles will starve. The severity and incidence of many plant diseases can also be minimized by crop rotation, and selection of the planting site may affect the severity of insect infestations.

Trap crops are planted to attract and hold pest insects where they can be managed more efficiently and prevent or reduce their movement onto valuable crops. Early planted potatoes can act as a trap crop for Colorado potato beetles emerging in the spring. Since the early potatoes are the only food source available, the beetles will congregate on these plants where they can more easily be controlled.

Adjusting the timing of planting or harvest is another cultural control technique. The earlier planted processing tomatoes grown in the western United States are far less likely to be infested by the tomato fruitworm than those planted later in the season. It is also important to use pest-free transplants. Some vegetable crop transplants can be infested with insect pests, and growers using these transplants are put at a considerable disadvantage.

PHYSICAL AND MECHANICAL CONTROL

The use of physical barriers such as row covers or trenches prevents insects from reaching the crop. Row covers can help prevent early-season damage to cucurbits by cucumber beetles, and plastic-lined trenches are effective in trapping large numbers of dispersing Colorado potato beetles in the spring and fall.

Cold storage is also considered a physical control and, although it does not necessarily kill the insect pests, it at least stops their development and further feeding on the stored crop. Other methods include hand picking of pests, sticky boards or tapes for control of flying insects in greenhouses and various trapping techniques.

Chapter 2

Pathogens Associated to the Seeds

As an introduction we will remind some aspects of the morphology and anatomy of the seed, in order to understand in a better way the effect pathogens have in the seed and later, the plant originated from it.

THE SEED

The seed consists of three basic parts:

- Embryo,
- Storage tissues
- Seed coat.

The embryo has a reproductive function, being capable of initiating cellular divisions and growth. It is the most important part of the seed. It is an axis that originates growth in two directions, with the objective of originating a root and a shoot. Usually, the embryo is very small, compared to other parts of the seed. It is like a miniature plant. It is consists of a radicle, plumule, one or two cotyledons, and hypocotyl or epicotyl, depending on the type of plant.

Energy storage is localized in the cotyledons, in the endosperm or the perisperm. Cotyledons originate from the zygote and are part of the embryo. In many species, storage substances are localized in the cotyledons, and the embryo develops absorbing all the endosperm.

Episperm is the seed cover, consists of two layers, the testa and the endopleura. The outer layer is the testa; it can be stony, leathery, membranous or fleshy. Over the testa we can recognize: the hilum, scar or point of attachment of the seed to the funiculus, water penetrates easily through it; Micropyle, the point upon the seed at which was the orifice of the ovule through which the pollen tube enters; raphe, suture originated by the part of the funiculus that is fused along the side of the ovule. The endopleura is the inner layer; it is thin and generally whitish. Teguments, testa or protective covers delimit the seed. They are formed by one or more layers of cells originated from the ovule integuments and sometimes from the pericarp made from the walls of the ovary.

Regarding the healthiness or pathology of the seed we will consider three aspects that allow us to understand processes associated to seed-borne diseases: infection of the seed, infection of the plant and steps that can be taken to reduce the damage caused by this relationship.

Seed Infection Mechanisms

The area of science that studies the relationship between pathogens and seeds is Seed Pathology. It does not only identify the pathogens, it also includes the role of the seed as source of inoculum, the survival of the pathogen and the actions taken to control the pathogens associated to it. It uses the knowledge of General Pathology, Microbiology and Seed Analysis. To obtain a perspective of seed borne diseases, seed borne microorganisms can be considered fewer than four classes.

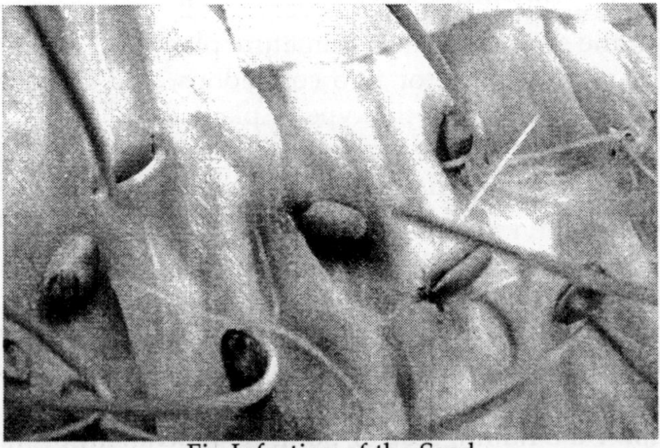

Fig.Infection of the Seed

The first consists of pathogens for which the seed is the main source of inoculum; when seed infection is controlled, the disease is controlled. An example would be lettuce mosaic virus. For these pathogens, the importance of seed-borne inoculum has long been recognized, and control practices have been developed. The second class consists of important pathogens in which the seed borne phase of the disease is of minor significance as a source of inoculum.

Examples are those in which the crop residues in the field were the major source of inoculum. The third and largest group of seed borne microorganisms consists of those that have never been shown to cause disease as a result of their presence on seeds. The fourth class is a group of microorganisms that can infect the seed either in the field or in storage and reduce yield and seed quality. Examples of field fungi are *Diplodia, Fusarium, Cladosporium,* etc.

The storage fungi *Aspergillus* and *Penicillium* can invade most types of seeds under high-moisture storage conditions. The process of seed infection is influenced by the conditions under which the crop grows.

Between the facts that influence in the process of infection are: the host and its genotype, the pathogen and its genotype end environmental facts. An infected seed will not always an infected seed will be the cause of the infection of the infection in the plant which originates, so, it is to say that the infection caused from an infected seed is an exception, not something to happen generally. There are two situations, Systemic infection of the seed and Contamination or Infestation of the seed.

Systemic Infection of the Seed

The establishment of a pathogen in any part of the seed is refereed as seed infection. It can be systemic, by the vascular system or plasmodesmata or directly by natural or artificial wounds. The same pathogen can infect the seed using one or more of these mechanisms.

For example: *Xanthomonas campestris* pv. *phaseoli* can infect seed through the vascular system, by natural openings (From the pod suture goes to the funiculus, then to the raphe and tegument, or it can also happen thought the micropyle).

Systemic Infection through Flowers, Fruits or Funiculus

Most of the systemic seed-borne bacteria and fungus reach and infect the embryo through the flower or from the peduncule of the fruit, via funiculus. Viruses go to the embryo from the systemically infected mother plant and the infected or contaminated pollen. They rarely reach the embryo during the formation of the seed or the embryo itself.

Examples of some infections that occur through the vascular system are: Some forma specialis of Fusarium oxysporum in pumpkin, pea and tomato; Plasmopara halstedii in sunflower; Septoria glycines in soy; Verticilium dahliae in spinach and sugar beet; Certain pathovares of Xanthomonas campestris in bean, cabbage, rice and sweet pepper; and Pseudomonas syringae pv. Lachymans in cucumber.

Penetration through the Stigma

During the infection, some pathogens follow the same path as the pollen grains do. Spores of some fungus reach the stigma and germinate, producing an hypha that reaches the ovary through the style, where they can stay as a dormant mycelium until seed germination.

For example: Ustilago nuda and U. triticci in barley and wheat, and Alternaria alternata in sweet pepper. Viruses can infect through infected pollen, the male gamete carries the virus, when joining the ovule it generates an infected embryo. If both, the male and female gamete are infected they can even produce an infected endosperm.

Penetration through the Wall of the Ovary

Some fungi, like *Ustilago nuda* and *U. triticci* penetrate through the wall of the ovary as a result of the germination of the Teliospores on the stigma or the wall of the ovary. The pro-mycelium goes through the wall and other tissues until it reaches the embryo. In some other cases, penetration occurs through breakages on the testa, establishing itself in the endopleura or the endosperm. In fleshy fruits, like cucumber, melon, eggplant, tomato, sweet pepper and others, contamination can occur directly through the funiculus or in the tegument, during the process of seed formation.

Examples of this are Colletotrichum lagenarium in watermelon and Rhizoctonia solani, when it invades fleshy fruits, it is capable of infecting from the placenta and penetrate to the developing ovule or seeds that are still in its formation process and have not lignified its cover.

Penetration through Wounds and Natural Openings

Natural openings like the hilum and the micropyle or wounds generated during the thresh are spots where pathogens like *Xanthomonas campestris* pv. *phaseolicola* in bean and *Pseudomonas syringae* pv. *lachrymans* in cucumber.

SEED CONTAMINATION OR INFESTATION

Contamination or infestation refers to the passive relationship of a pathogens and seeds. The pathogen itself or parts of it can stick or can get mixed with the seeds during any of the processes during seed recollection: harvesting, extraction, thresh, selection and packing.

Pathogens that Stick to the Surface of the Seed

Pathogens that stick to seeds during harvest or postharvest do so by their spores (Clamidospores, Oospores, Teliospores, Uredospores), bacterial cells and in some cases, virions. Spores of the following fungi can be carried on seed coat surfaces: Alternaria brassicae and A. brassicicola in crucifers; A. longipes in tobacco; A. radicina in carrot; Ascochyta pinodella in pea; Drechslera sorokiniana and D. oryzae in rice; D. avenae in oats; Fusarium oxysporum f. sp. callistephi in China aster; Sclerotia of Rhizoctonia solani in eggplant, pepper, and tomato; Tilletia caries, T. foetida, T. contraversa, and Urocystis agropyri in wheat.

A number of bacteria, such as the following, contaminate seed surfaces: Corynebacterium flaccumfacjens pv. flaccumfacjens in bean, Pseudomonas syringae pv. phaseolicola in bean, P. syringae pv. lachrymans in cucumber,

P. syringae pv. tomato in tomato, C. rnichiganense pv. michiganense in pepper, *Xanthomonas campestris* pv. *campestris* in cabbage Also some viruses as Tobacco mosaic virus, Tomato mosaic virus, Pepper Mosaic virus.

ACCOMPANYING CONTAMINATION

This type of contamination refers to physical mixing of the seed with pathogen's propagation organs like the sclerotium, nematode's galls, contaminated plant parts or soil particles containing pathogens.

Structures of the Pathogens

Some fungi produce resistance structures named sclerotiums that are made of compacted mycelium. Under certain conditions of temperature and humidity it can germinate, it is common that this occurs together with the seed, producing its infection, by direct penetration or by the spores generated by the fructification organs (Apothecium or Perithecium). These spores are carried by the wind and taken to susceptible tissues of the plant, generally, some parts of the flower. Sclerotiums have many shapes and during thresh can be easily included with the rest of the seed.

Some examples of this are Sclerotinia sclerotiorum in horticultural, grain and flower crops, Sclerotinia cepivorum in garlic and Claviceps purpurea in barley.

Mix with infected Plant Parts

Infected plant's parts and residues can carry fructifications or spores of fungi and bacteria. This situation is particularly common in cereal, forage and oily crop's seeds. Examples of this are Septoria nodorum in wheat, Puccinia malvacearum in hollyhock (malvarrosa), Colletotrichum trifolii in clover, Erwinia carotovora pv. carotovora in tobacco, Pseudomonas syringae pv. phaseolicola in bean and Sclerotinia sclerotiorum in sunflower, soy and peanut.

Soil

Seeds can be mixed with contaminated soil and, carry

micro-sclerotiums of *Macrophomina phaseolina* in dirty seeds of kidney bean, Verticillium oxysporum f.sp. phaseoli in cotton seeds, Plasmodiophora brassicae in turnip and Fusarium oxysporum f. sp. phaseoli in bean.

SEED TRANSMITION

We saw how the seed gets infected or infested. Now we will see how these seeds can produce an ill plant. This is to say, how the inoculum goes from the seed to the plant. We call seed-borne pathogens only those that can produce an infection. Because they have to be distinguished from those that can associate to the seed but do not produce an infection, these ones are called pathogens not transmitted by seed.

The virus that causes curly top in sugar beet can be present in the perisperm but does not cause an infection in the plant. In general terms, infection can be classified into systemic and non systemic. It is systemic when the pathogen introduces itself to the plant when the seed germinates, and develops with it. Non systemic infection occurs when there is a localized infection caused by the pathogen in the seedling at the stage of pre or post emergency, in this situation there are no systemic symptoms.

Systemic Transmition

This type of infection can be produced by pathogens that are carried with the seed in various parts, like the embryo, endosperm or episperm, or by contamination of the outer portion of this one.

Infection of the Embryo

When the seed germinates, if the embryo is infected, the pathogen initiates its growth together with the plant. Symptoms can show up during different stages of development. In the case of *Ustilago triticci,* the mycelium grows together with the plant and expresses symptoms only at the flowering stage, which are that all the tissues of the spike, exept the rachis, are replaced by spores.

Some pathovares of *Xanthomonas campestris* that infect

cabbage, bean or sweet pepper move between the cells of the host until they reach the vascular system and produce symptoms in leaves or stems. Most of seed-borne viruses persist inside the embryo. Its multiplication and movement accompanies the plant during its development, and there can be symptoms at any stage, from the formation of first leaves until flowering or fructification.

Non Embryo Infection

Infection of the episperm (testa and endopleura) occasionally conducts to a systemic infection. Some bacteria, like Corynebacterium michiganense pv. michiganense in tomato and Xanthomonas campestris pv. campestris in cabbage, penetrate through stomata of cotyledons, and from there, reach the vascular system, initiating the systemic infection.

Episperm Contamination

In some few cases this type of contamination conducts to a systemic infection. Most of these exceptions are fungi that are highly specialized in their pathogenesis and produce in cereals the so called smuts, rusts or mildews.

In these cases, generally, spores are carried outside the seed, they germinate, penetrate the coleoptile, and start a systemic infection. This can occur directly o by a more complex system of haploid hypha fusion in genera like Tilletia, Ustilago, Uromyces, Sclerospora, Pernospora and Puccinia.

Non Systemic Transmition

Non systemic infection is very common, and in the same way as systemic infection, it can come from an infection, outside contamination or by pathogens mixed with the seeds.

Infection of the Embryo

This case is restricted to some pathogenic fungi that maintain itself in the embryo or the episperm as hypha inside the seed. Primary infection starts as injuries in the cotyledons or primary leaves, stems or petiole. Fungi fructifications

(Pycnidium, Acervulus) can develop on these organs, under certain favorable conditions (temperature and humidity).

These fructifications produce spores that, with the action of water and wind, disperse the disease to other parts of the plant and other plants. This occurs in *Ascochyta pisi* in pea, *Colletotrichum lindemuthianum* in bean and C. *truncatum* in soy.

Infection of the Episperm

Generally, seeds in which the episperm is infected, do not geminate, or germinate and contaminate the soil. Rarely produce a systemic infection, but can infect the seedling from the outside. Over the injuries, new inoculum is produced; this one infects other parts of the plant or other plants. *Septoria nodorum* in wheat under high humidity conditions forms Picnidium in the coleoptile, whose spores disseminate the disease to other plants.

Contamination of the Episperm

Contamination outside the seed's testa can produce healthy seedlings, but the inoculum infects the soil and from there it can cause infections at more developed stages of the plant. Wheat and rice grains that are contaminated with Teliospores of Neovossia indica in wheat and N. torrida in rice, produce Esporidisporas in the soil, that can be carried by the wind and infect flowers, originating the smuts.

Accompanying Contamination

The mixture of seeds with sclerotiums of the fungi or contaminated soil particles produces non systemic infections, at any stage of the plant growth and development. It is common for *Sclerotinia sclerotiorum* to produce mycelium from the sclerotiums. These mycelium can directly infect the plant or produce fructifications that produce spores that can be carried by the wind and infect flowers of different species, like sunflower, soy, peanut, etc.

Prevention of Seedborne Diseases

Prevention of seed-borne diseases is based in two main

facts. On one hand there is the usage of material free from pathogens and contaminants that could originate a disease and on the other there are the treatments to eliminate the inoculum in the seed or propagule. When using material free from pathogens it is important to have adequate pathogen detection methods, this is to say, sensible and specific. No production of health quality seed can be done without valid methods to detect the possibility of contamination. Some treatments can be necessary to eliminate the pathogens from the seeds, or at leafs, maintain an economically viable balance.

Health Testing of the Seed

One of the most relevant aspects about controlled health quality seed is refereed to the detection methods used on every case. ISTA (International Seed Testing Association) appeared in 1924, by was only interested in seed's gene purity and germination aspects. But detection methods were applied.

According to the criterion of each laboratory, so their results were not comparative. So, in 1957, PDC established a comparative health testing programme and standardize applied methods. The complexity for comparison of these methods made a considerable step back on the publication of working sheets (protocols). ISTA approved methods became Rules. By 1999 only 64 protocols were being compared, and only 14 of them were accepted as Rules.

As the solutions proposed by ISTA did not satisfy the needs of international seed industry, this industry, in 1994, organized The International Seed Health Initiative for Vegetables (ISHI-Veg) chartered by the vegetable seed industries in The Netherlands and France. The Initiative was soon joined by the seed companies in the United States, Israel and Japan. This group represents the production of over 75% of the worlds vegetable seed supply. ISHI published many protocols, organized by crop. For horticultural crops, the ISHI-VEG, the Seed Health Testing Methods

ISHI-Veg Status

- An ISHI Validated Reference Method is a method

that has been through the ISTA/ISHI comparative testing process and is being published as ISTA wn ISHI Reference Method is a method that has been through ISHI comparative testing and is under review by the ISTA/ISHI reviewers for publication as ISTA Working Sheet.

- An ISHI Accepted Method is a method that is commonly used by the seed industry for determining seed health. The method has been published in a journal, as a working sheet and/or is publicly available for comparative testing and commonly in use by the seed industry. The method is in comparative testing or aspects of the test are being tested for inclusion or deletion from the method.
- An ISHI Reviewed Method is a method that is publicly available or published, documented and prioritized by the ITG, but not subjected to a comparative test at this time. The method usually is accepted by other agencies or groups.

In 1995, during the second ISTA-PDC symposium it was decided that a Joint ISTA/ISHI Guidelines for Comparative Testing of Methods for Detection of Seed Borne Pathogens would be edited.

Through this combined effort many laboratories, organized under this alignments, the ISTA developed the Manual of Seed Health Testing Methods that has two sections:

- Validated Test Methods
- Peer Reviewed Methods.

The transition process from working protocol to ruled method is long and laborious, so working protocols can be considered as adequate methods for seed analysis. Nowadays, it is common that technological advances may become obsolete even before they are published.

The protocol to analyse *Xanthomonas campesrtis* pv.*campestris*, which causes black rot in crucifers, was published in 1982, there, it was recommended one minute to extract the bacteria from the seeds to starch culture medium. 2 years later, this time was raised to 2 hours, and other specific culture mediums. In 1987 the protocol was replaced, again in 1996 and at present time again it is under revision.

Minimum requirements to Produce Certified Seed

The Organization for Economic Cooperation and Development (OECD) establishes steps to be taken in order to produce basic and certified seed. Between these steps it is emphasized that generations from the mother material to the basic material are strictly limited in number. The number of certified seed generations, after the basic seed, varies from crop to crop and technical conditions.

In both, sexually and vegetative reproduced species, the schemes include similar steps. It generally starts with selection of a plant with genetic and health advantages. If health condition was not satisfying, certain actions should be taken in order to release it from considered pathogens. This Mother Plant is maintained under strict special security conditions in order to maintain healthiness (isolated glasshouses or chambers).

In the Certification Schemes two particularities abut seed borne pathogens should be considered, those transmitted exclusively with the seed and those that can also be transmitted with contaminated soil and plant rests. Diseases transmitted exclusively with the seed can be controlled by using clean healthy seed.

This seed can be obtained by Seed Certification Programmes or by treatments that are really effective, like *in vitro* culture, thermotherapy, chemotherapy or a combined action of these, for systemic pathogens.

Tolerance levels in certified seed depend on the multiplication speed and dispersion of the inoculum of the pathogen of the considered disease. If a small amount of inoculum may be epidemic, like Anthracnose and the bean bacterial blight, if levels above zero chemical control may not be enough to stop the disease.

But pathogens whose inoculum multiplies slowly generation after generation, the disease may be evident only several generations later, so, if contamination levels are above zero may be accepted in adequate control methods are applied to the seed. This is what happens in wheat, with the Loose Smut, continuous treatments diminish inoculum levels, and make it perfectly tolerable.

New Techniques for Seed Health Analysis

Pathogen detection methods have suffered great changes since 1957. Back then it was mainly targeted to fungi, and based exclusively on incubation and identification, or the cultivation of these seeds and counting the percentage of unhealthy seedlings. These analyses were very laborious and required a high infection level in order to be detected.

Nowadays, technological advances have allowed detecting even low infection levels, evaluating 1,000 to 10,000 seeds at once. Specificity of culture mediums, usage of antibiotics and other products has allowed isolating the pathogen from the seed.

Together with this, the use of reactives as mono or polyclonal antiserums and molecular markers is also available. The Lettuce Mosaic Virus (LMV), back in 1950 the symptoms were observed over many thousands of seedlings from the seed that were going to be analyzed.

Later, technical advances determined that 30,000 seedlings were needed, and even later, in 1983, an alternative to the *Chenopodium quinoa* test was discovered, the ELISA test.

Watermelon Fruit Blotch caused by bacteria is a severe problem in glasshouse winter production. It was determined that a single contaminated seed in 9,000 of them was enough too widespread the bacteria to the rest of the plants. That is the reason why screening was done over 10,000 seedlings. Nowadays PCR technique is used.

Chapter 3

Plants and Poor Soil

WHEN a soil is sick, either because its beneficial bacteria do not perform their functions properly, or because of abnormalities in its chemical or physical properties, careful treatment and proper cultural methods may restore it to health. But when a soil becomes sick and unproductive because parasitic forms gain a foothold in it, much greater skill and knowledge are required to cope with the problem. Its solution is of the greatest economic importance to the gardener and to the greenhouse man.

Parasitic fungi, upon finding their way into a soil, do not necessarily interfere with the work of the beneficial bacteria, such as the ammonifiers and nitrifiers, for instance. Nor do they always influence the chemical or physical nature of the soil. Many of them directly attack the crop itself, causing serious diseases in the plants.

DAMPING OFF

This disease is very familiar to every grower of plants. It is peculiar to seedlings or tender plants, and is very prevalent in the greenhouse, the hot bed, the cold frame as well as in the field. It is induced by the presence of definite parasitic fungi, which thrive best in overwatered soils, and when the greenhouse is kept at a comparatively high temperature with poor ventilation. Damping off is also favoured by thick sowing and too much shade in the seed bed.

Symptoms of Damping Off. Every experienced grower knows the disease when he sees it. Seedlings freshly damped off are soft and water soaked at the base of the stem. If they are pulled they often break off easily. A more careful examination shows that the root system is entirely decayed, although the upper part of the stem and leaves may still be green, and also possibly fresh.

The degree of prostration in the seedlings is determined by the amount of moisture in the soil. If it is slight, the seedlings will become flabby and wilted before they topple over. With a high moisture content, they are more firm, but become prostrate as soon as infection sets in. The trouble usually begins in spots in the bed, thence spreading in every direction. Damping off is usually caused by several fungi, the chief of which is Pythium de Baryanum Hesse.

The organism was first named and described by Hesse in 1874. Ward found it to be a very prevalent parasite in the garden soils of Europe. In America the fungus was first recognized as of great economic importance by Atkinson. The seedlings of most greenhouse plants may become subject to damping off by Pythium.

When examined under a compound microscope, Pythium de Baryanum is seen to be made up of coarse non-separate, highly granular, irregularly branched hyaline vegetative threads or mycelium. The younger growing threads are more finely granular. The oldest are coarsely granular or more often empty. These threads penetrate the cells of the host, where they obtain its food.

Pythium de Baryanum does not often fruit on the dead seedlings. The fruiting is better observed when the fungus is grown in pure culture. Under normal conditions it produces two forms of spores, conidia and oogonia. The summer spores, or conidia, are swellings formed at the tip of the hyphæ. These swellings readily break off from the mother threads and germinate by sending out a slender tube. This tube penetrates the seedling tissue where it grows and develops and after due incubation reproduces the disease.

The oospore, or sexual spore, is the stage which is most commonly found. The female organ (oogonium) first develops as a terminal enlargement which is cut off by a septum from the mother thread. Next or adjacent to it a slender tube is cut off from the mycelium by a septum. This tube performs the function of the male sexual organ and is known as antheridium. The latter then comes into close contact and empties all its content into the female oogonium. Fertilization thus takes place, and a mature egg, or oospore, or winter resting spore is formed.

The latest investigations have not yet disclosed whether or not Pythium de Baryanum is carried over from year to year by its oospores. It is apparently able to propagate itself indefinitely by its vegetative mycelium.

Of the other fungi which are capable of producing a damping off in the greenhouse or seed bed may be mentioned Sclerotinia libertiana Fckl., Phoma solani Halst., Colletotrichum sp., Fusarium sp., Sclerotium rolfsii Sacc. and Rhizactonia solani Kuhn. Each of these, except the last, will be taken up separately in connection with the study of their respective hosts.

The fungi which cause damping off are introduced into the greenhouse, primarily with sick soil used in the compost, and also with infected manure. The practice of dumping diseased plants and all other infected material in the manure pile cannot be too strongly condemned. Sometimes very lightly infected plants with no visible symptoms of disease in the seed bed may nevertheless act as carriers, and like-wise infect the greenhouse soil.

DAMPING OFF OF CUTTINGS

Greenhouse men are often troubled with a damping off of cuttings. In specific cases this is brought about by parasitic fungi which, however, will be taken up at length under the discussion of the various hosts.

ROOT ROT

Although not so virulent as Pythium, Rhizoctonia is a frequent cause of considerable failure in green-house culture. The fungus causes a damping off of seedlings and cuttings and a serious root rot.

Symptoms. The symptoms of Rhizoctonia rot or wilt do

not differ materially from those produced by Pythium de Baryanum. On older plants, however, Rhizoctonia produces cankers or deep lesions which are very characteristic. These are formed on the roots as well as on the base of the stem. The lesions are reddish brown and extend into the cortical or vital layer as well as into the woody tissue.

Fig. Root Rot

There is perhaps no other parasitic fungus which is so widespread and which is capable of attacking such a variety of hosts as Rhizoctonia. The work of Peltier shows that the following greenhouse crops are susceptible to Rhizoctonia: beet, bean, cauliflower, celery, cucumber, egg plant, horseradish, lettuce, muskmelon, pepper, radish, tomato, sweet alyssum, amaranthus, ornamental asparagus, china aster, begonia, candytuft, carnation, coleus, dianthus, lavatera, lobelia, pansy, poinsettia, sweet pea, violet.

Cuttings of the following hosts are also reported by Peltier to damp off from Rhizoctonia: Abutilon hybridum, var. lavitzii, Acalypha wilkesiana, var. bicolour, A. wilkesiana, var. tricolour; A. wilkesiana, var. marginata, Ageratum mexicanum, Alyssum odoratum; Coleus, Cuphea phatycentra, Tresine, Petunia, Piguerua trinervia, Lautolina chamoecyparissus, Sedum spectabile, Althernanthera, Vincia major.

The Organism. In the United States the first extended account of Rhizoctonia was given by Pammel. Many other excellent accounts by American workers have appeared from time to time, to which we shall have occasion to refer later.

The genus Rhizoctonia includes several forms of sterile fungi, all of which are distinguished by the manner of growth in pure culture, and by its mycelium. Young hyphae of R. solani Kuhn are at first hyaline, then deepening in colour from a yellowish to a deep brown. The young branches are somewhat narrowed at their point of union with the parent hyphæ and grow in a direction almost parallel with each other. A septum is also laid down at a short distance from the point of union with the parent mycelium.

There is another form of hypha which is made up of barrel shaped cells each of which is capable of germinating like a spore. In pure cultures R. solani produces sclerotia which are first soft, whitish, and which later become hard and dark. The fungus is, carried over from year to year as sclerotia which are able to withstand the effects of heat, cold, drought, or moisture.

Next in importance to Rhizoctonia is a group of fungi which belong to the genus Fusarium. Soils infected with these species of fungi become unfit for tomatoes, sweet peas, etc., thereby causing great financial losses to the greenhouse man. Individual difficulties will be taken up in studying each of these crops separately. As an illustration of a typical Fusarium sick soil let us consider the wilt of sweet pea. The cause of this trouble is a soil inhabiting fungus, Fusarium lath yri Taub.

Symptoms. The first symptom of the disease is a sudden flagging of the leaves, accompanied by general wilting and collapse of the seedling. Usually upon sowing the seeds a fair percentage germinate and reach the height of about 8 to 10 inches before they are attacked by the fungus. If the collapsed seedlings are allowed to remain on the ground, the stems will soon be covered with the sickle shaped spores. Eventually the decayed tissue rots and is soon invaded by small fruit flies which now begin to distribute the fungus from place to place by carrying its spores.

The Organism. The mycelium of Fusarium lath yri is hyaline, septate and branched. At an early age the mycelial cells round up into countless numbers of chlamydospores. Old cultures are practically one mass of these resting bodies. The spores are of two sorts, the macroconidia which are sickle shaped, 3-4 septate, the microconidia are one celled, minute spherical to elliptical.

ROOT KNOT

Caused by Heterodera radicicola (Greef) Mull. Although root knot is most prevalent in light soils, it may sometimes be found in heavier lands. The trouble is most widespread out of doors in the South-ern States, where the winter is mild. In the North the worm is usually unable to winter over in the open unless it is protected by trash or dead weeds. It is, however, prevalent in greenhouses and is undoubtedly introduced with sick soil brought in from the field.

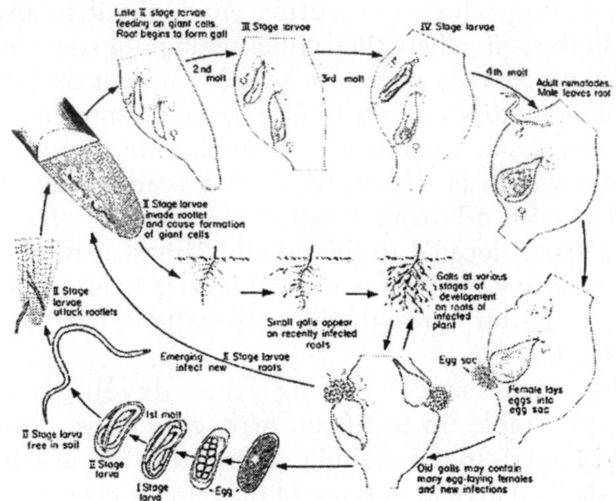

Fig.Root Knot

Symptoms. The disease is characterized by swellings or knots on the roots. These swellings may be variously shaped, and are often mistaken for the true nodules of legumes. Infected plants become stunted, pale, and usually linger for a long time before dying.

The Organism. The nematode is a very minute worm, seldom exceeding one twenty-fifth of an inch in length. It is semi-transparent so that it cannot be easily detected by the naked eye. In searching for the eel worm, it is necessary to break a fresh knot. Close examination will reveal two types of worms; a spindle shaped worm, the male, and a pearly white pear shaped organism, the female, firmly embedded in the gall tissue. The female is very prolific, depositing no less than 400 to goo eggs during her lifetime.

The eggs are whitish, semi-transparent, bean shaped bodies, and too small to be noticed without the aid of a magnifying glass. The time which elapses until the eggs hatch depends largely upon weather conditions. In warm days the eggs hatch sooner than in cold days.

Upon hatching, the young larvæ either remain in the tissue of the host plant in which they emerge, or, as is more often the case, leave the host and enter the soil. This is the only period during which the worms move about to any great extent in the soil, where they either remain for some length of time or immediately penetrate an-other root of the host.

The nematodes in most cases become completely buried in the root tissue, establishing themselves in the soft cellular structure which is rich in food. The head of the worm is provided with a boring apparatus consisting of a sharply pointed spear, located in the mouth. This structure not only aids it in getting food but is also valuable in helping the young worms to batter through the cell walls before becoming definitely located.

The two sexes during the development are indistinguishable up to fifteen or twenty days, both being spindle shaped. In the molting or shedding of the skin, there is a marked change in the case of the female, especially in the posterior region of the body, which no longer possesses a tail-like appendage. Fertilization occurs soon after this molt, and many radical changes occur in the shape and structure of the organization of the worm.

The fertilized female increases rapidly in breadth and becomes a pearly white flask or pear-shaped individual. At this stage it is far from being worm-like and may, therefore, be overlooked by one unfamiliar with the life-history of the eel worm. The young male is much like that of the young female larvæ, being spindle shaped in outline.

The male does not cause as much damage to the root tissue as the female, and its purpose in life seems to be only that of fertilizing the female, for after this function has been performed, it is quite probable that the male worm takes no more food.

Omnivorous Nature of the Eel Worm. There are two hundred and thirty-five species of plants known to suffer from the eel worm. This number includes all the important families of the flowering plants. According to Bessey * the following are among the greenhouse plants subject to root knot: bean, beet, cantaloupe, cauliflower, cucumber, egg-plant, lettuce, radish, tomato. For methods of control, see p. 40.

LEAF BLIGHT NEMATODE

Beside the root knot disease which is caused by Heterodera radicicola there is another nematode which confines its injury to foliage only. Of the greenhouse hosts affected by this pest may be mentioned the Begonia, Asplenium nidus-avis, Pteris serrulata avistata, Pteris wimeseth, Pteris tremula, and Pelargonium.

Symptoms. On the Cincinnati begonia the symptoms, according to Clinton, are manifested as numerous small indistinct discolourations limited by the small veinlets. In time, however, the tiny spots enlarge and unite, forming a conspicuous reddish-brown blotch. Frequently infection is manifested as long streaks along the rnain veins. Often isolated spots occur in the midst of the surrounding healthy tissue.

On Asplenium nidus-avis, the trouble becomes conspicuous in dark brown areas from the base of the leaf near the midrib. These spread up-ward until the entire lower half of the leaf is killed. On Pteris, the spots appear as reddish brown bands reaching out from the midrib to the border, but limited sidewise by the small parallel cross veins.

The Organism. The nematode in question is a slender

microscopical worm. The latter chooses the air chamber of the leaf in which to lay its eggs and upon hatching travels around in different parts of the same leaf or to the neighbouring foliage. The worm can travel only when there is a wet film on the leaves.

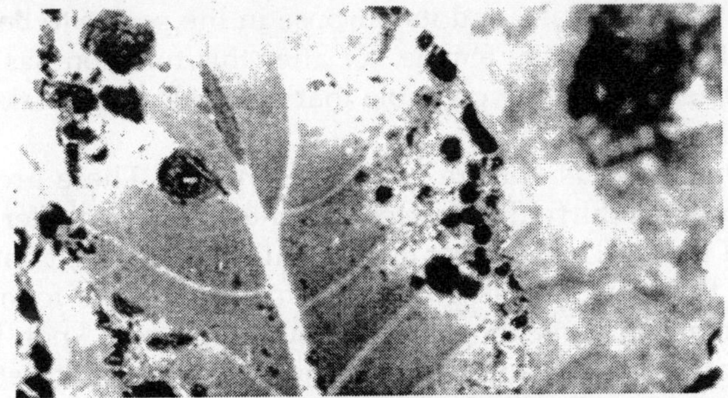

Fig. Leaf Blight Nematode

Control. Immersing fern plants for five minutes in water heated to 122 degrees F. (50 degrees C.) does not seem to injure the ferns, but seems to kill the nematode. All infected leaves should be cut off and burned. Diseased plants should be isolated from healthy ones. Spraying with Bordeaux may also act as a repellant.

SOIL

Soil, the loose material that covers the land surfaces of Earth and supports the growth of plants. In general, soil is an unconsolidated, or loose, combination of inorganic and organic materials. The inorganic components of soil are principally the products of rocks and minerals that have been gradually broken down by weather, chemical action, and other natural processes. The organic materials are composed of debris from plants and from the decomposition of the many tiny life forms that inhabit the soil.

Soils vary widely from place to place. Many factors determine the chemical composition and physical structure of the soil at any given location. The different kinds of rocks, minerals, and other geologic materials from which the soil originally formed play a role. The kinds of plants or other vegetation that grow in the soil are also important. Topography—that is, whether the terrain is steep, flat, or some combination—is another factor. In some cases, human activity such as farming or building has caused disruption. Soils also differ in colour, texture, chemical makeup, and the kinds of plants they can support.

Soil actually constitutes a living system, combining with air, water, and sunlight to sustain plant life. The essential process of photosynthesis, in which plants convert sunlight into energy, depends on exchanges that take place within the soil. Plants, in turn, serve as a vital part of the food chain for living things, including humans. Without soil there would be no vegetation—no crops for food, no forests, flowers, or grasslands. To a great extent, life on Earth depends on soil.

The study of different soil types and their properties is called soil science or pedology. Soil science plays a key role in agriculture, helping farmers to select and support the crops on their land and to maintain fertile, healthy ground for planting. Understanding soil is also important in engineering and construction. Soil engineers carry out detailed analysis of the soil prior to building roads, houses, industrial and retail complexes, and other structures.

Soil takes a great deal of time to develop—thousands or even millions of years. As such, it is effectively a nonrenewable resource. Yet even now, in many areas of the world, soil is under siege. Deforestation, over-development, and pollution from humanmade chemicals are just a few of the consequences of human activity and carelessness. As the human population grows, its demand for food from crops increases, making soil conservation crucial.

Quality of Soil

The good or bad qualities of a soil have reference to the needs of the crops which are to be grown upon it, and it is only after a consideration of the requirements of plants that a clear conception can be formed of what characters the soil must possess for it to be a suitable medium on which healthy crops can be raised.

In the first place, soil, to be of any use, must be sufficiently loose and porous to allow the roots of plants to grow and extend freely. It may be so compact that root development is checked or stopped altogether, in which case the plant suffers. On the other hand it should not be too open in texture or the roots do not get a proper hold of the ground and are easily disturbed by wind: moreover such soils are liable to blow away, leaving the underground parts exposed to the air and drought.

The roots like all other parts of plants contain protoplasm or living material, which cannot carry on its functions unless it is supplied with an adequate amount of oxygen: hence the necessity for the continuous circulation of fresh air through the soil.

If the latter is too compact or has its interstices filled with carbon dioxide gas or with water - as is the case when the ground is water-logged - the roots rapidly die of suffocation just as would an animal under the same conditions.

There is another point which requires attention. Plants need very considerable amounts of water for their nutrition and growth; the waterholding capacity is, therefore, important. If the soil holds too much it becomes water-logged and its temperature falls below the point for healthy growth, at any rate of the kinds of plants. Usually cultivated on farms and in gardens. If it allows of too free drainage drought sets in and the plants, not getting enough water for their needs, become stunted in size. Too much water is bad, and too little is equally injurious.

In addition, the temperature of the soil largely controls the yield of crops which can be obtained from the land. Soil whose temperature remains low, whether from its northerly aspect or from its high water content or other cause, is unsatisfactory, because the germination of seeds and the general life processes of plants cannot go on satisfactorily except at certain temperatures well above freezing-point.

A good soil should be deep to allow of extensive root development and, in the case of arable soils, easy to work with implements. Even when all the conditions above mentioned in regard to texture, water-holding capacity, aeration and temperature are suitably fulfilled the soil may still be barren: plant foodmaterial is needed. This is usually present in abundance although it may not be available to the plant under certain circumstances, or may need to be replenished or increased by additions to the soil of manures or fertilizers.

CHIEF CONSTITUENTS OF THE SOIL

An examination of the soil shows it to be composed of a vast number of small particles of sand, clay, chalk and humus, in which are generally imbedded larger or smaller stones. It will be useful to consider the nature of the four chief constituents just mentioned and their bearing upon the texture, water-holding capacity and other characters which were referred to in the previous section.

Sand consists of grains of quartz or flint, the individual particles of which are large enough to be seen with the unaided eye or readily felt as gritty grains when rubbed between the finger and thumb. When a little soil is shaken up with water in a tumbler the sand particles rapidly fall to the bottom and form a layer which resembles ordinary sand of the seashore or river banks.

Chemically pure sand is silicon dioxide (SiO_2) or quartz, a clear transparent glass-like mineral, but as ordinarily met with, it is more or less impure and generally coloured reddish or yellowish by oxide of iron.

A soil consisting of sand entirely would be very loose, would have little capacity to retain water, would be liable to become very hot in the daytime and cool at night and would be quite unsuitable for growth of plants.

The term clay is often used by chemists to denote hydrated silicate of alumina ($Al_2O_3.2SiO_2.2H_2O$), of which kaolin or china clay is a fairly pure form. This substance is present in practically all soils but in comparatively small amounts. Even in the soils which farmers speak of as stiff clays it is rarely present to the extent of more than I or 2%.

The word " clay " used in the agricultural sense denotes a sticky intractable material which is found to consist of exceedingly fine particles (generally less than. 005 mm. in diameter) of sand and other minerals derived from the decomposition of rocks, with a small amount of silicate of alumina.

The peculiar character which clay possesses is probably due not to its chemical composition but to its physical state. When wet it becomes sticky and almost impossible to move or work with farm implements; neither air nor water can penetrate freely. In a dry state it becomes hard and bakes to a brick. It holds water well and is consequently cold, needing the application of much heat to raise its temperature. It is obvious, therefore, that soil composed entirely of clay is as useless as pure sand so far as the growth of crops upon it is concerned.

Chalk consists, when quite pure, of calcium carbonate ($CaC03$), a white solid substance useful in small amounts as a plant foodmaterial, though in excess detrimental to growth. Alone, even when broken up into small pieces, it is unsuitable for the growth of plants.

Humus, the remaining constituent of soil, is the term used for the decaying vegetable and animal matter in the soil. A good illustration of it is peat. Its water-holding capacity is great, but it is often acid, and when dr y it is light and incapable of supporting the roots of plants properly. Few of the commonly cultivated crops can live in a soil consisting mainly of humus.

From the above account it will be understood that not one of the four chief soil constituents is in itself of value for the growth of crops, yet when they are mixed, as they usually are in the soils met with in nature, one corrects the deficiencies of the other. A perfect soil would be such a blend of sand, clay, chalk and humus as would contain sufficient clay and humus to prevent drought, enough sand to render it pervious to fresh air and prevent waterlogging, chalk enough to correct the tendency to acidity of the humus present, and would have within it various substances which would serve as food-materials to the crops.

COMPOSITION OF SOILS

Soils comprise a mixture of inorganic and organic

components: minerals, air, water, and plant and animal material. Mineral and organic particles generally compose roughly 50 percent of a soil's volume. The other 50 percent consists of pores—open areas of various shapes and sizes. Networks of pores hold water within the soil and also provide a means of water transport. Oxygen and other gases move through pore spaces in soil. Pores also serve as passageways for small animals and provide room for the growth of plant roots.

Inorganic Material

The mineral component of soil is made up of an arrangement of particles that are less than 2.0 mm (0.08in) in diameter. Soil scientists divide soil particles, also known as soil separates, into three main size groups: sand, silt, and clay.

According to the classification scheme used by the United States Department of Agriculture (USDA), the size designations are: sand, 0.05 to 2.00 mm (0.002 to 0.08 in); silt 0.002 to 0.05 mm (0.00008 to 0.002 in); and clay, less than 0.002 mm (0.00008 in). Depending upon the rock materials from which they were derived, these assorted mineral particles ultimately release the chemicals on which plants depend for survival, such as potassium, calcium, magnesium, phosphorus, sulfur, iron, and manganese.

Organic Material

Organic materials constitute another essential component of soils. Some of this material comes from the residue of plants, the remains of plant roots deep within the soil, or materials that fall on the ground, such as leaves on a forest floor. These materials become part of a cycle of decomposition and decay, a cycle that provides important nutrients to the soil. In general, soil fertility depends on a high content of organic materials.

Even a small area of soil holds a universe of living things, ranging in size from the fairly large to the microscopic: earthworms, mites, millipedes, centipedes, grubs, termites, lice, springtails, and more. And even a gram of soil might contain as many as a billion microbes—bacteria and fungi too small to be seen with the naked eye.

All these living things form a complex chain: Larger creatures eat organic debris and excrete waste into the soil, predators consume living prey, and microbes feed on the bodies of dead animals. Bacteria and fungi, in particular, digest the complex organic compounds that make up living matter and reduce them to simpler compounds that plants can use for food. A typical example of bacterial action is the formation of ammonia from animal and vegetable proteins.

Other bacteria oxidize the ammonia to form nitrogen compounds called nitrites, and still other bacteria act on the nitrites to form nitrates, another type of nitrogen compound that can be used by plants. Some types of bacteria are able to fix, or extract, nitrogen directly from the air and make it available in the soil.Ultimately, the decay of plant and animal material results in the formation of a dark-coloured organic matter known as humus. Humus, unlike plant residues, is generally resistant to further decomposition.

CHEMICAL COMPOSITION OF THE SOIL

It has been found by experiment that plants need for their nutritive process and their growth, certain chemical elements, namely, carbon, hydrogen, oxygen, nitrogen, sulphur, phosphorus, potassium, magnesium, calcium and iron. With the exception of the carbon and a small proportion of the oxygen and nitrogen, which may be partially derived from the air, these elements are taken from the soil by crops.

The following table shows the amounts of the chief constituents removed by certain crops in lb per acre: - Plants also remove from the soil silicon, sodium, chlorine, and other elements which are, nevertheless, found to be unessential for the growth and may therefore be neglected here.

Leguminous crops take some of the nitrogen which they require from the air, but most plants obtain it from the nitrates present in the soil. The sulphur exists in the soil chiefly in the form of sulphates of magnesium, calcium and other metals; the phosphorus mainly as phosphates of calcium, magnesium and iron; the potash, soda and other bases as silicates and nitrates; calcium and magnesium carbonates are also common constituents of many soils. In the ordinary chemical analyses of the soil determinations are made of the nitrogen and various carbonates present as well as of the amount of phosphoric acid, potash, soda, magnesia and other components soluble in strong hydrochloric acid.

Soil Formation

Soil formation is an ongoing process that proceeds through the combined effects of five soil-forming factors: parent material, climate, living organisms, topography, and time. Each combination of the five factors produces a unique type of soil that can be identified by its characteristic layers, called horizons. Soil formation is also known as *pedogenesis* (from the Greek words *pedon,* for "ground," and *genesis,* meaning "birth" or "origin").

Parent Material

The first step in pedogenesis is the formation of parent material from which the soil itself forms. Roughly 99 percent of the world's soils derive from mineral-based parent materials that are the result of weathering, the physical disintegration and chemical decomposition of exposed bedrock. The small percentage of remaining soils derives from organic parent materials, which are the product of environments where organic matter accumulates faster than it decomposes. This accumulation can occur in marshes, bogs, and wetlands.

Bedrock itself does not directly give rise to soil. Rather, the gradual weathering of bedrock, through physical and chemical processes, produces a layer of rock debris called regolith. Further weathering of this debris, leading to increasingly smaller and finer particles, ultimately results in the creation of soil.In some instances, the weathering of bedrock creates parent materials that remain in one place. In other cases, rock materials are transported far from their source—blown by wind, carried by moving water, and borne inside glaciers.

Climate

Climate directly affects soil formation. Water, ice, wind,

heat, and cold cause physical weathering by loosening and breaking up rocks. Water in rock crevices expands when it freezes, causing the rocks to crack.

Rocks are worn down by water and wind and ground to bits by the slow movement of glaciers. Climate also determines the speed at which parent materials undergo chemical weathering, a process in which existing minerals are broken down into new mineral components. Chemical weathering is fastest in hot, moist climates and slowest in cold, dry climates.

Climate also influences the developing soil by determining the types of plant growth that occur. Low rainfall or recurring drought often discourage the growth of trees but allow the growth of grass. Soils that develop in cool rainy areas suited to pines and other needle-leaf trees are low in humus.

Living Organisms

As the parent material accumulates, living things gradually gain a foothold in it. The arrival of living organisms marks the beginning of the formation of true soil. Mosses, lichens, and lower plant forms appear first. As they die, their remains add to the developing soil until a thin layer of humus is built up. Animals' waste materials add nutrients that are used by plants. Higher forms of plants are eventually able to establish themselves as more and more humus accumulates. The presence of humus in the upper layers of a soil is important because humus contains large amounts of the elements needed by plants.

Living organisms also contribute to the development of soils in other ways. Plants build soils by catching dust from volcanoes and deserts, and plants' growing roots break up rocks and stir the developing soil. Animals also mix soils by tunneling in them.

WATER IN THE SOIL

The importance of an adequate supply of water to growing crops cannot well be over-estimated. During the life of a plant there is a continuous stream of water passing through it which enters by the root-hairs in the soil and after passing along the stem is given off from the 'stomata of the leaves into the open air above ground. It has been estimated that an acre of cabbage will absorb from the land and transpire from its leaves more than ten tons of water per day when the weather is fine.

In addition to its usefulness in maintaining a turgid state of the young cells without which growth cannot proceed, water is itself a plant food-material and as absorbed from the soil contains dissolved in it all the mineral food constituents needed by plants for healthy nutrition. Without a sufficient supply plants remain stunted and the crop yield is seriously reduced, as we see in dry seasons when the rainfall is much below the average.

If one condition is more necessary than another for good crops it is a suitable supply of water, for no amount of manuring or other treatment of the soil will make up for a deficient rainfall. The amount needed for the most satisfactory nutrition varies with different plants. In the case of fair average farm crops it has been shown that for the production of one ton of dry matter contained in them from 300 to 500 tons of water has been absorbed and utilized by the plants. This may be more than the rainfall, in which case irrigation or special control of the water supply may be necessary.

The water-holding capacity of a soil depends upon the amount of free space between the particles of which it is composed into which water can enter. In most cases this amounts to from 30 to 50% of the volume of the soil.

When the pore-space of the soil is filled with water it becomes water-logged and few plants can effect absorption by their roots under such conditions. The root-hairs die from want of air, and the whole plant soon suffers. Fields of wheat and other cereals rarely recover after a week's submergence, but orchards and many trees when at rest in winter withstand a flooded or water-logged condition of the soil for two or three weeks without damage.

The most satisfactory growth is maintained when the amount of water present is not more than 40 to 60% of what would saturate it. Under such conditions each particle of soil is surrounded by a thin film of water and in the pore-space air can freely circulate. It is from such films that the root-hairs absorb all that plants require for their growth.

The movement of water into the root-hairs is brought about by the osmotic action of certain salts in their cell-sap. Crops are, however, unable to absorb all the water present in the soil, for when the films become very thin they are held more firmly or cling with more force to the soil particles and resist the osmotic action of the root-hairs. Plants have been found to wither and die in sandy soils containing i a % of water, and in clay soils in which there was still present 8% of water.

When a long glass tube open at both ends is filled with soil and one end is dipped in a shallow basin of water, the water is found to move upwards through the soil column just as oil will rise in an ordinary lamp wick. By this capillary action water may be transferred to the upper layers of the soil from a depth of several feet below the surface. In this manner plants whose roots descend but a little way in the ground are enabled to draw on deep supplies.

Not only does water move upwards, but it is transferred by capillarity in all directions through the soil. The amount and speed of movement of water by this means, and the distance to which it may be carried, depend largely upon the fineness of the particles composing the soil and the spaces left between each. The ascent of water is most rapid through coarse sands, but the height to which it will rise is comparatively small. In clays whose particles are exceedingly minute the water travels very slowly but may ultimately reach a height of many feet above the level of the " water-table " below.

While this capillary movement of water is of great importance in supplying the needs of plants it has its disadvantages, since water may be transferred to the surface of the soil, where it evapo rates into the air and is lost to the land or the crop growing upon it..

The loss in this manner was found to be in one instance over a pound. of water per day per square foot of surface, the " water-table " being about 4 or 5 ft. below.

One of the most effective means of conserving soil moisture is by " mulching," *i.e.* by covering the surface of the soil with some loosely compacted material such as straw, leaf-refuse or stablemanure. The space between the parts of such substances is too large to admit of capillary action; hence the water conveyed to the surface of the soil is prevented from passing upwards any further except by slow evaporation through the mulching layer.

A loose layer of earth spread over the surface of the soil acts in the same way, and a similarly effective mulch may be prepared by hoeing the soil, or stirring it to a depth of one or two inches with harrows or other implements. The hoe and harrow are therefore excellent tools for use in dry weather. Rolling the land is beneficial to young crops. in dry weather, since it promotes capillary action by reducing the soil spaces. It should, however, be followed by a light hoeing or harrowing.

In the semi-arid regions of the United States, Argentina and other countries where the average annual rainfall lies between ioa to 20 in., irrigation is necessary to obtain full crops every year. Good crops, however, can often be grown in such areas without irrigation if attention is paid to the proper circulation of water in the soil and means for retaining it or preventing excessive loss by evaporation. Of course care must be exercised in the selection of plants - such as sorghum, maize, wheat, and alfalfa or lucerne - which are adapted to dry conditions and a warm climate.

So far as the water-supply is concerned - and this is what ultimately determines the yield of crops - the rain which falls upon the soil should be made to enter it and percolate rapidly through its interstices. A deep porous bed in the upper layers is essential, and this should consist of fine particles which lie close to each other without any tendency to stick together and " puddle " after heavy showers. Every effort should be made to prepare a good mealy tilth by suitable ploughing, harrowing and consolidation.

In the operation of ploughing the furrow slice is separated from the soil below, and although in humid soils this layer may be left to settle by degrees, in semi-arid regions this loosened layer becomes. dry if left alone even for a few hours and valuable water evaporates. into the air. To prevent this various implements, such as disk harrows and specially constructed rollers, may be used to consolidate the upper stirred portion of the soil and place it in close capillary relationship with the lower unmoved layer.

If the soil is allowed to become dry and pulverized, rain is likely to run off or " puddle the surface without penetrating it more than a very short distance. Constant hoeing or harrowing to maintain a natural soil mulch layer of 2 or 3 in. deep greatly conserves the soil water below. In certain districts where the rainfall is low a crop can only be obtained once every alternate year, the intervening season being devoted to tillage with a view of getting the rain into the soil and retaining it there for the crop in the following year.

Soil scientists also characterize soils according to how effectively they retain and transport water. Once water enters the soil from rain or irrigation, gravity comes into play, causing water to trickle downward. Water is also taken up in great quantities by the roots of plants: Plants use anywhere from 200 to 1,000 kg (440 to 2,200 lb) of water in the formation of 1 kg (2.2 lb) of dry matter. Soils differ in their capacity to retain moisture against the pull exerted by gravity and by plant roots. Coarse soils, such as those consisting of mostly of sand, tend to hold less water than do soils with finer textures, such as those with a greater proportion of clays.

Water also moves through soil pores independently of gravity. This movement can occur via capillary action, in which water molecules move because they are more attracted to the pore walls than to one another. Such movement tends to occur from wetter to drier areas of the soil. The movement from soil to plant roots can also depend on how tightly water molecules are bound to soil particles.

The attraction of water molecules to each other is an example of cohesion. The attraction of water molecules to other materials, such as soil or plant roots, is a type of adhesion. These effects, which determine the so-called matric potential of the soil, depend largely on the size and arrangement of the soil particles. Another factor that can affect water movement is referred to as the osmotic potential. The osmotic potential hinges on the amount of dissolved salts in the soil. Soils high in soluble salt tend to reduce uptake of water by plant roots and seeds. The sum of the matric and osmotic potentials is called the total water potential. In soil, water carries out the essential function of bringing mineral nutrients to plants. But the balance between water and air in the soil can be delicate. An overabundance of water will saturate the soil and fill pore spaces needed for the transport of oxygen.

The resulting oxygen deficiency can kill plants. Fertile soils permit an exchange between plants and the atmosphere, as oxygen diffuses into the soil and is used by roots for respiration. In turn, the resulting carbon dioxide diffuses through pore spaces and returns to the atmosphere.

This exchange is most efficient in soils with a high degree of porosity. For farmers, gardeners, landscapers, and others with a professional interest in soil health, the process of aeration—making holes in the soil surface to permit the exchange of air—is a crucial activity. The burrowing of earthworms and other soil inhabitants provides a natural and beneficial form of aeration.

Bacteria in the Soil.

Recent science has made much progress in the investigation of the micro-organisms of the soil. Whereas the soil used to be looked upon solely as a dead, inert material containing certain chemical substances which serve as food constituents. of the crops grown upon it, it is now known to be a place of habitation for myriads of minute living organisms upon whose activity much of its fertility depends.

They are responsible for many important chemical processes which make the soil constituents more available and better adapted to the nutrition of crops. One cubic centimetre of soil taken within a foot or so from the surface contains from II to 2 millions of bacteria of many different kinds, as well as large numbers of fungi. In the lower depths of the soil the numbers. decrease, few being met with at a depth of 5 or 6 ft.

The efficiency of many substances, such as farm-yard manure,. guanos, bone-meal and all other organic materials, which are spread over or dug or ploughed into the land for the benefit of farm and' garden crops, is bound up with the action of these minute living beings.

Without their aid most manures would be useless for plant growth. Farm-yard manure, guanos and other fertilizers. undergo decomposition in the soil and become broken down into compounds of simple chemical composition better suited for absorption by the roots of crops, the changes involved being directly due to the activity of bacteria and fungi. Much of the work carried on by these organisms is not clearly understood; there are, however, certain processes which have been extensively investigated and to these it is necessary to refer.

It has been found by experiment that the nitrogen needed by practically all farm crops except leguminous ones is best supplied in the form of a nitrate; the rapid effect of nitrate of soda when used' as a top dressing to wheat or other plants is well known to farmers.

It has long been known that when organic materials such as the dung and urine of animals, or even the bodies of animals and plants, are applied to the soil, the nitrogen within them becomes oxidized, and ultimately appears in the form of nitrate of lime, potash or some other base.

The nitrogen in decaying roots, in the dead stems. and leaves of plants, and in humus generally is sooner or later changed into a nitrate, the change being effected by bacteria. That: the action of living organisms is the cause of the production of nitrates is supported by the fact that the change does not occur when the soil is heated nor when it is treated with disinfectants which destroy or check the growth and life of bacteria.

The process resulting in the formation of nitrates in the soil is spoken of as nitrification. The steps in the breaking down of the highly complex nitrogenous proteid compounds contained in the humus of the soil, or applied to the latter by the farmer in the form of dung and organic refuse generally, are many and varied; most frequently the insoluble proteids are changed by various kinds of putrefactive bacteria into soluble proteids (peptones, &c.), these into simpler amido-bodies, and these again sooner or later into compounds 7of ammonia.

The urea in urine is also rapidly converted by the uro-bacteria into ammonium carbonate. The compounds of ammonia thus formed from the complex substances by many varied kinds of micro-organisms are ultimately oxidized into nitrates.

The change takes place in two stages and is effected by two special groups of nitrifying bacteria, which are present in all soils. In the first stage the ammonium compounds are oxidized to nitrites by the agency of very minute motile bacteria belonging to the genus Nitrosomonas. The further oxidation of the nitrite to a nitrate is effected by bacteria belonging to the genus Nitrobacter.

Several conditions must be fulfilled before nitrification can occur. In the first place an adequate temperature is essential; at 5° or 6° C. (40-43° F.) the process is stopped, so that it does not go on in winter. In summer, when the temperature is about 24° C. (75° F.), nitrification proceeds at a rapid rate.

The organisms do not carry on their work in soils deficient in air; hence the process is checked in water-logged soils. The presence of a base such as lime or magnesia (or their carbonates) is also essential, as well as an adequate degree of moisture: in dry soils nitrification ceases.

It is the business of the farmer and gardener to promote the activity of these organisms by good tillage, careful drainage and occasional application of lime to soils which are deficient in this substance. It is only when these conditions are attended to that decay and nitrification of dung, guano, fish-meal, sulphate of ammonia and other manures take place, and the constituents which they contain become available to the crops for whose benefit they have been applied to the land.

Nitrates are very soluble in water and are therefore liable to be washed out of the soil by heavy rain. They are, however, very readily absorbed by growing plants, so that in summer, when nitrification is most active, the nitrates produced are usually made use of by crops before loss by drainage takes place. In winter, however, and in fallows loss takes place in the subsoil water.

There is also another possible source of loss of nitrates through the activity of denitrifying bacteria. These organisms reduce nitrates tö nitrites and finally to ammonia and gaseous free nitrogen which escapes into the atmosphere. Many bacteria are known which are capable of denitrification, some of them being abundant in fresh dung and upon old straw. They can, however, only carry on their work extensively under anaerobic conditions, as in waterlogged soils or in those which are badly tilled, so that there is but little loss of nitrates through their agency.

An important group of soil organisms are now known which have the power of using the free nitrogen of the atmosphere for the formation of the complex nitrogenous compounds of which their bodies are largely composed. By their continued action the soil becomes enriched with nitrogenous material which eventually through the nitrification process becomes available to ordinary green crops. This power of " fixing nitrogen," as it is termed, is apparently not possessed by higher green plants.

The bacterium, *Clostridium pasteurianum*, common in most soils, is able to utilize free nitrogen under anaerobic conditions, and an organism known as *Azotobacter chroococcum* and some others closely allied to it, have similar powers which they can exercise under aerobic conditions. For the carrying on of their functions they all need to be supplied with carbohydrates or other carbon compounds which they obtain ordinarily from humus and plant residues in the soil, or possibly in some instances from carbohydrates manufactured by minute green algae with which they live in close union.

Certain bacteria of the nitrogenfixing class enter into association with the roots of green plants, the best-known examples being those which are met with in the nodules upon the roots of clover, peas, beans, sainfoin and other plants belonging to the leguminous order.

That the fertility of land used for the growth of wheat is improved by growing upon it a crop of beans or clover has been long recognized by farmers. The knowledge of the cause, however, is due to modern investigations. When wheat, barley, turnips and similar plants are grown, the soil upon which they are cultivated becomes depleted of its nitrogen; yet after a crop of clover or other leguminous plants the soil is found to be richer in nitrogen than it was before the crop was grown. This is due to the nitrogenous root residues left in the land. Upon the roots of leguminous plants characteristic swollen nodules or tubercles are present.

These are found to contain large numbers of a bacterium termed *Bacillus radicicola* or *Pseudomonas radicicola.* The bacteria, which are present in almost all soils, enter the root-hairs of their host plants and ultimately stimulate the production of an excrescent nodule, in which they live. For a time after entry they multiply, obtaining the nitrogen necessary for their nutrition and growth from the free nitrogen of the air, the carbohydrate required being supplied by the pea or clover plant in whose tissues they make a home.

The nodules increase in size, and analysis shows that they are exceedingly rich in nitrogen up to the time of flowering of the host plant. During this period the bacteria multiply and most of them assume a peculiar thickened or branched form, in which state they are spoken of as bacteroids.

Later the nitrogen-content of the nodule decreases, most of the organisms, which are largely composed of proteid material, becoming digested and transformed into soluble nitrogenous compounds which are conducted to the developing roots and seeds. After the decay of the roots some of the unchanged bacteria are left in the soil, where they remain ready to infect a new leguminous crop.

The nitrogen-fixing nodule bacteria can be cultivated on artificial media, and many attempts have been made to utilize them for practical purposes. Pure cultures may be made and after dilution in water or other liquid can be mixed with soil to be ultimately spread over the land which is to be infected.

The method of using them most frequently adopted

consists in applying them to the seeds of leguminous plants before sowing, the seed being dipped for a time in a liquid containing the bacteria. In this manner organisms obtained from red clover can be grown and applied to the seed of red clover; and similar inoculation can be arranged for other species, so that an application of the bacteria most suited to the particular crop to be cultivated can be assured.

In many cases it has been found that inoculation, whether of the soil or of the seed, has not made any appreciable difference to the growth of the crop, a result no doubt due to the fact that the soil had already contained within it an abundant supply of suitable organisms.

But in other instances greatly increased yields have been obtained where inoculation has been practised. More or less pure cultures of the nitrogen-fixing bacteria belonging to the *Azotobacter* group have been tried and recommended for application to poor land in order to provide a cheap supply of nitrogen.

The application of pure cultures of bacteria for improving the fertility of the land is still in an experimental stage. There is little doubt, however, that in the near future means will be devised to obtain the most efficient work from these minute organisms, either by special artificial cultivation and subsequent application to the soil, or by improved methods of encouraging their healthy growth and activity in the land where they already exist.

Improvement of Soils

The fertility of a soil is dependent upon a number of factors, some of which, such as the addition of fertilizers or manures, increase the stock of available food materials in the soil while others, such as application of clay or humus, chiefly influence the fertility of the land by improving its physical texture.

The chief processes for the improvement of soils are:liming, claying and marling, warping, paring and burning, and green manuring. Most of these more or less directly improve the land by adding to it certain plant food constituents which are lacking, but the effect of each process is in reality very complex. In the majority of cases the good results obtained are more particularly due to the setting free of " dormant " or " latent " food constituents and to the amelioration of the texture of the soil, so that its aeration, drainage, temperature and water-holding capacity are altered for the better.

The material which chemists call calcium carbonate is met with in a comparatively pure state in chalk. It is present in variable amounts in limestones of all kinds, although its white ness may there be masked by the presence of iron oxide and other coloured substances. Carbonate of lime is also a constituent to a greater or lesser extent in almost all soils. In certain sandy soils and in a few stiff clays it may amount to less than 4%, while in others in limestone and chalk districts there may be 50 to 80% present. Pure carbonate of lime when heated loses 44% of its weight, the decrease being due to the loss of carbon dioxide gas.

The resulting white product is termed calcium oxide lime, burnt lime, quicklime, cob lime, or caustic lime. This substance absorbs and combines with water very greedily, at the same time becoming very hot, and falling into a fine dry powder,' calcium hydroxide or slaked lime, which when left in the open slowly combines with the carbon dioxide of the air and becomes calcium carbonate, from which we began.

When recommendations are made about liming land it is necessary to indicate more precisely than is usually done which of the three classes of material named above - chalk, quicklime or slaked lime - is intended. Generally speaking the oxide or quicklime has a more rapid and greater effect in modifying the soil than slaked lime, and this again greater than the carbonate or chalk.

Lime in whatever form it is applied has a many-sided influence in the fertility of the land. It tends to improve the tilth and the capillarity of the soil by binding sands together somewhat and by opening up clays. If applied in too great an amount to light soils and peat land it may do much damage by rendering them too loose and open.

The addition of small quantities of lime, especially in a caustic form, to stiff greasy clays makes them much more porous and pliable. A lump of clay, which if dried would become hard and intractable, crumbles into pieces when dried after adding to it 2% of lime. The lime causes the minute separate particles of clay to flocculate or group themselves together into larger compound grains between which air and water can percolate more freely.

It is this power of creating a more crumbly tilth on stiff clays that makes lime so valuable to the farmer. Lime also assists in the decomposition of the organic matter or humus in the soil and promotes nitrification; hence it is of great value after green manuring or where the land contains much humus from the addition of bulky manures such as farm-yard dung.

This tendency to destroy organic matter makes the repeated application of lime a pernicious practice, especially on land which contains little humus to begin with. The more or less dormant nitrogen and other constituents of the humus are made immediately available to the succeeding crop, but the capital of the soil is rapidly reduced, and unless the loss is replaced by the addition of more manures the land may become sterile. Although good crops may follow the application of lime, the latter is not a direct fertilizer or manure and is no substitute for such. Its best use is obtained on land in good condition, but not where the soil is poor.

When used on light dry land it tends to make the land drier, since it destroys the humus which so largely assists in keeping water in the soil. Lime is a base and neutralizes the acid materials present in badly drained meadows and boggy pastures. Weeds, therefore, which need sour conditions for development are checked by liming and the better grasses and clovers are encouraged. It also sets free potash and possibly other useful plant food-constituents of the soil. Liming tends to produce earlier crops and destroys the fungus which causes finger-and-toe or club-root among turnips and cabbages.

Land which contains less than about z /c, of lime usually needs the addition of this material. The particular form in which lime should be applied for the best results depends upon the nature of the soil. In practice the proximity to chalk pits or lime kilns, the cost of the lime and cartage, will determine which is most economical.

Generally speaking light poor lands deficient in organic matter will need the less caustic form or chalk, while quicklime will be most satisfactory on the stiff clays and richer soils. On the stiff soils overl y ing the chalk it was formerly the custom to dig pits through the soil to the rock below. Shafts 20 or 30 ft. deep were then sunk, and the chalk taken from horizontal tunnels was brought to the surface and spread on the land at the rate of about 60 loads per acre. Chalk should be applied in autumn, so that it may be split by the action of frost during the winter.

Quicklime is best applied, perhaps, in spring at the rate of one ton per acre every six or eight years, or in larger doses- 4 to 8 tons - every 15 to 20 years. Small dressings applied at short intervals give the most satisfactory results. The quicklime should be placed in small heaps and covered with soil if possible until it is slacked and the lumps have fallen into powder, after which it may be spread and harrowed in. Experiments have shown that excellent effects can be obtained by applying 5 or 6 cwt. of ground quicklime.

Gas-lime is a product obtained from gasworks where quicklime is used to purify the gas from sulphur compounds and other objectionable materials. It contains a certain amount of unaltered caustic lime and slacked lime, along with sulphates and sulphides of lime, some of which have an evil odour. As some of these sulphur compounds have a poisonous effect on plants, gas-lime cannot be applied to land directly without great risk or rendering it incapable of growing crops of any sort - even weeds - for some time.

It should therefore be kept a year or more in heaps in some waste corner and turned over once or twice so that the air can gain access to it and oxidize the poisonous ingredients in it. Many soils of a light sandy or gravelly or peaty nature and liable to drought and looseness of texture can be improved by the addition of large amounts of clay of an ordinary character.

Similarly soils can be improved by applying to them marl, a substance consisting of a mixture of clay with variable proportions of lime. Some of the chalk marls, which are usually of a yellowish or dirty grey colour, contain clay and 50 to 80% of carbonate of lime with a certain proportion of phosphate of lime. Such a material would not only have an influence on the texture of the land but the lime would reduce the sourness of the land and the phosphate of lime supply one of the most valuable of plant foodconstituents.

The beneficial effects of marls may also be partially due to the presence in them of available potash. Typical clay-marls are tenacious, soapy clays of yellowish-red or brownish colour and generally contain less than 50% of lime. When dry they crumble into small pieces which can be readily mixed with the soil by ploughing.

Many other kinds of marls are described; some are of a sandy nature, others stony or full of the remains of small shells. The amount and nature of the clay or marl to be added to the soil will depend largely upon the original composition of the latter, the lighter sands and gravel requiring more clay than those of firmer texture. Even stiff soils deficient in lime are greatly improved in fertility by the addition of marls.

In some cases as little as 40 loads per acre have been used with benefit, in others 180 loads have not been too much. The material is dug from neighbouring pits or sometimes from the fields which are to be improved, and applied in autumn and winter. When dry and in a crumbly state it is harrowed and spread and finally ploughed in and mixed with the soil.

On some of the strongest land it was formerly the practice to add to and plough into it burnt clay, with the object of making the land work more easily. The burnt clay moreover carried *Cl ay* with it potash and other materials in a state readily available to the crops. The clay is dug from the land or from ditches or pits and placed in heaps of 60 to ioo loads each, with faggot wood, refuse coals or other fuel.

Great care is necessary to prevent the heaps from becoming too hot, in which case the clay becomes baked into hard lumps of brick-like material which cannot be broken up.

With careful management, however, the clay dries and bakes, becoming slowly converted into lumps which readily crumble into a fine powder, in which state it is spread over and worked into the land at the rate of 40 loads per acre.

The paring and burning of land, although formerly practised as an ordinary means of improving the texture and fertility of arable fields, can now only be looked upon as a practice p to be adopted for the purpose of bringing rapidly into cultivation very foul leys or, land covered with a coarse turf. The practice is confined to poorer types of land, such as heaths covered with furze and bracken or fens and clay areas smothered with rank grasses and sedges.

To reduce such land to a fit state for the growth of arable crops is very difficult and slow without resort to paring and burning. The operation consists of paring off the tough sward to a depth of I to 2 in. just sufficient to effectually damage the roots of the plants forming the sward and then, after drying the sods and burning them, spreading the charred material and ashes over the land.

The turf is taken off either with the breast plough - a paring tool pushed forward from the breast or thighs by the workman - or with specially constructed paring ploughs or shims. The depth of the sod removed should not be too thick or burning is difficult and too much humus is destroyed unnecessarily, nor should it be too thin or the roots of the herbage are not effectually destroyed.

The operation is best carried out in spring and summer. After being pared off the turf is allowed to dry for a fortnight or so and is then placed in small heaps a yard or two wide at the base, a little straw or wood being put in the middle of each heap, which is then lighted.

As burning proceeds more turf is added to the outside of the heaps in such a manner as to allow little access of air. Every care should be taken to burn and char the sod thoroughly without permitting the heap to blaze. The ashes should be spread as soon as possible and covered by a shallow ploughing. The land is then usually sown with some rapidly growing green crop, such as rape, or with turnips.

Paring and burning improves the texture of clay lands, particularly if draining is carried out at the same time. It tends to destroy insects and weeds, and gets rid of acidity of the soil. No operation brings old turf into cultivation so rapidly. Moreover the beneficial effects are seen in the first crop and last for many years.

Many of the mineral plant food-constituents locked up in the coarse herbage and in the upper layers of the soil are made immediately available to crops. The chief disadvantage is the loss of nitrogen which it entails, this element being given off into the air in a free gaseous state.

It is best adapted for application to clays and fen lands and should not be practised on shallow light sands or gravelly soils, since the humus so necessary for the fertility of such areas is reduced too much and the soil rendered too porous and liable to suffer from drought.

Many thousands of acres of low-lying peaty and sandy land adjoining the tidal rivers which flow into the Humber have been improved by a process termed " warping." The warp consists of fine muddy sediment which is suspended in the tidal river water and appears to be derived from material scoured from the bed of the Humber by the action of the tide and a certain amount of sediment brought down by the tributary streams which join the Humber some distance from its mouth.

The field or area to be warped must lie below the level of the water in the river at high tide. It is first surrounded by an embankment, after which the water from the river is allowed to flow through a properly constructed sluice in its bank, along a drain or ditch to the land which is prepared for warping. By a system of carefully laid channels the water flows gently over the land, and deposits its warp with an even level surface.

At the ebb of the tide the more or less clear water flows back again from the land into the main river with sufficient force to clean out any deposit which may have accumulated in the drain leading to the warped area, thus allowing free access of more warpladen water at the next tide. In this manner poor peats and sands may be covered with a large layer of rock soil capable of growing excellent crops.

The amount of deposit laid over the land reaches a thickness of two or three feet in one season of warping, which is usually practised between March and October, advantage being taken of the spring tides during these months.

The new warp is allowed to lie fallow during the winter after being laid out in four-yard " lands " and becomes dry enough to be sown with oats and grass and clover seeds in the following spring. The clover-grass ley is then grazed for a year or two with sheep, after which wheat and potatoes are the chief crops grown on the land.

Green manures are crops which are grown especially for the purpose of ploughing into the land in a green or actively growing state. The crop during its growth obtains a considerable amount of *Green.* carbon from the carbon dioxide of the air, and builds it up *Manuring. i* nto compounds which when ploughed into the land become humus.

The carbon compounds of the latter are of no direct nutritive value to the succeeding crop, but the decaying vegetable tissues very greatly assist in retaining moisture in light sandy soils, and in clay soils also have a beneficial effect in rendering them more open and allowing of better drainage of superfluous water and good circulation of fresh air within them. The ploughing-in of green crops is in many respects like the addition of farm-yard manure.

Their growth makes no new addition of mineral food-constituents to the land, but they bring useful substances from the subsoil nearer to the surface, and after the decay of the buried vegetation these become available to succeeding crops of wheat or other plants. Moreover, where deep-rooting plants are grown the subsoil is aerated and rendered more open and suitable for the development of future crops.

The plants most frequently used are white mustard, rape, buckwheat, spurry, rye, and several kinds of leguminous plants, especially vetches, lupins and serradella. By far the most satisfactory crops as green manures are those of the leguminous class, since they add to the land considerable amounts of the valuable fertilizing constituent, nitrogen, which is obtained from the atmosphere.

By nitrification this substance rapidly becomes available to succeeding crops. On the light, poor sands of Saxony Herr Schultz, of Lupitz, made use of serradella, yellow lupins and vetches as green manures for enriching the land in humus and nitrogen, and found the addition of potash salts and phosphates very profitable for the subsequent growth of potatoes and wheat. He estimated that by using leguminous crops in this manner for the purpose of obtaining cheap nitrogen he reduced the cost of production of wheat more than 50%.

The growing crops should be ploughed in before flowering occurs; they should not be buried deeply, since decay and nitrification take place most rapidly and satisfactorily when there is free access of air to the decaying material. When the crop is luxuriant it is necessary to put a roller over it first, to facilitate proper burial by the plough.

The best time for the operation appears to be late summer and autumn. (J. PE.) *Soil and Disease. - The* influence of different kinds of soil as a factor in the production of disease requires to be considered, in regard not only to the nature and number of the microorganisms they contain, but also to the amount of moisture and air in them and their capacity for heat.

The moisture in soil is derived from two sources—the rain and the ground-water.

Above the level of the ground-water the soil is kept moist by capillary attraction and by evaporation of the water below, by rainfall, and by movements of the ground-water; on the other hand, the upper layers are constantly losing moisture by evaporation from the surface and through vegetation. When the ground-water rises it forces air out of the soil; when it falls again it leaves the soil moist and full of air.

The nature of the soil will largely influence the amount of moisture which it will take up or retain. In regard to water, all soils have two actions - namely, permeability and absorbability. Permeability is practically identical with the speed at which percolation takes place; through clay it is slow, but increases in rapidity through marls, loams, limestones, chalks, coarse gravels and fine sands, reaching a maximum in soil saturated with moisture.

The amount of moisture retained depends mainly upon the absorbability of the soil, and as it depends largely on capillary action it varies with the coarseness or fineness of the pores of the soil, being greater for soils which consist of fine particles. The results of many analyses show that the capacity of soils for moisture increases with the amount of organic substances present; decomposition appears to be most active when the moisture is about 4%, but can continue when it is as low as 2%, while it appears to be retarded by any excess over 4%.

Above the level of the ground-water all soils contain air, varying in amount with the degree of looseness of the soil. Some sands contain as much as 50% of air of nearly the same composition as atmospheric air. The oxygen, however, decreases with the depth, while the carbon dioxide increases.

SOIL CHARACTERISTICS

Scientists can learn a lot about a soil's composition and origin by examining various features of the soil. Colour, texture, aggregation, porosity, ion content, and pH are all important soil characteristics.

Colour

Soils come in a wide range of colours—shades of brown, red, orange, yellow, gray, and even blue or green. Colour alone does not affect a soil, but it is often a reliable indicator of other soil properties. In the surface soil horizons, a dark colour usually indicates the presence of organic matter.

Soils with significant organic material content appear dark brown or black. The most common soil hues are in the red-to-yellow range, getting their colour from iron oxide minerals coating soil particles. Red iron oxides dominate highly weathered soils. Soils frequently saturated by water appear gray, blue, or green because the minerals that give them the red and yellow colours have been leached away.

Texture

A soil's texture depends on its content of the three main

mineral components of the soil: sand, silt, and clay. Texture is the relative percentage of each particle size in a soil. Texture differences can affect many other physical and chemical properties and are therefore important in measures such as soil productivity.

Soils with predominantly large particles tend to drain quickly and have lower fertility. Very fine-textured soils may be poorly drained, tend to become waterlogged, and are therefore not well-suited for agriculture. Soils with a medium texture and a relatively even proportion of all particle sizes are most versatile. A combination of 10 to 20 percent clay, along with sand and silt in roughly equal amounts, and a good quantity of organic materials, is considered an ideal mixture for productive soil.

Aggregation

Individual soil particles tend to be bound together into larger units referred to as aggregates or soil peds. Aggregation occurs as a result of complex chemical forces acting on small soil components or when organisms and organic matter in soil act as glue binding particles together.

Soil aggregates form soil structure, defined by the shape, size, and strength of the aggregates. There are three main soil shapes: platelike, in which the aggregates are flat and mostly horizontal; prismlike, meaning greater in vertical than in horizontal dimension; and blocklike, roughly equal in horizontal and vertical dimensions and either angular or rounded. Soil peds range in size from very fine—less than 1 mm (0.04 in)—to very coarse—greater than 10 mm (0.4 in). The measure of strength or grade refers to the stability of the structural unit and is ranked as weak, moderate, or strong. Very young or sandy soils may have no discernible structure.

Porosity

The part of the soil that is not solid is made up of pores of various sizes and shapes—sometimes small and separate, sometimes consisting of continuous tubes. Soil scientists refer to the size, number, and arrangement of these pores as the soil's porosity. Porosity greatly affects water movement and gas exchange. Well-aggregated soils have numerous pores, which are important for organisms that live in the soil and require water and oxygen to survive. The transport of nutrients and contaminants will also be affected by soil structure and porosity.

Ion Content

Soils also have key chemical characteristics. The surfaces of certain soil particles, particularly the clays, hold groupings of atoms known as ions. These ions carry a negative charge. Like magnets, these negative ions (called anions) attract positive ions (called cations).

Cations, including those from calcium, magnesium, and potassium, then become attached to the soil particles, in a process known as cation exchange. The chemical reactions in cation exchange make it possible for calcium and the other elements to be changed into water-soluble forms that plants can use for food. Therefore, a soil's cation exchange capacity is an important measure of its fertility.

pH

Another important chemical measure is soil pH, which refers to the soil's acidity or alkalinity. This property hinges on the concentration of hydrogen ions in solution. A greater concentration of hydrogen results in a lower pH, meaning greater acidity. Scientists consider pure water, with a pH of 7, neutral. The pH of a soil will often determine whether certain plants can be grown successfully. Blueberry plants, require acidic soils with a pH of roughly 4 to 4.5. Alfalfa and many grasses, on the other hand, require a neutral or slightly alkaline soil. In agriculture, farmers add limestone to acid soils to neutralize them.

SOIL USE

For most of human history, soil has not been treated as the valuable and essentially nonrenewable resource that it is. Erosion has devastated soils worldwide as a result of overuse and misuse. In recent years, however, farmers and agricultural experts have become increasingly concerned with soil management.

Erosion

Erosion is the wearing away of material on the surface of the land by wind, water, or gravity. In nature, erosion occurs very slowly, as natural weathering and geologic processes remove rock, parent material, or soil from the land surface. Human activity, on the other hand, greatly increases the rate of erosion. In the United States, the farming of crops accounts for the loss of over 3 billion metric tons of soil each year.

In a cultivated field from which crops have been harvested, the soil is often left bare, without protection from the elements, particularly water. Raindrops smash into the soil, dislodging soil particles. Water then carries these particles away. This movement may take the form of broad overland flows known as sheet erosion.

More often, the eroding soil is concentrated into small channels, or rills, producing so-called rill erosion. Gravity intensifies water erosion. Landslides, in which large masses of water-loosened soil slide down an incline, are a particularly extreme example.

Wind erosion occurs where soils are dry, bare, and exposed to winds. Very small soil particles can be suspended in the air and carried away with the wind. Larger particles bounce along the ground in a process called saltation.

Soil Management

To prevent exposure of bare soil, farmers can use techniques such as leaving crop residue in the soil after harvesting or planting temporary growths, such as grasses, to protect the soil from rain between crop-growing seasons. Farmers can also control water runoff by planting crops along the slope of a hill (on the contour) instead of in rows that go up and down.

Soil faces many threats throughout the world. Deforestation, overgrazing by livestock, and agricultural practices that fail to conserve soil are three main causes of accelerated soil loss. Other acts of human carelessness also damage soil. These include pollution from agricultural pesticides, chemical spills, liquid and solid wastes, and acidification from the fall of acid rain.

Loss of green spaces, such as grassland and forested areas, in favour of impermeable surfaces, such as pavement, buildings, and developed land, reduces the amount of soil and increases pressure on what soil remains. Soil is also compacted by heavy machinery and off-road vehicles. Compaction rearranges soil particles, increasing the density of the soil and reducing porosity. Crusts form on compacted soils, preventing water movement into the soil and increasing runoff and erosion.

With the world's population now numbering upwards of 6 billion people—a figure that may rise to 10 billion or more within three decades—humans will depend more than ever on soil for the growth of food crops.

Yet the rapidly increasing population, the intensity of agriculture, and the replacement of soil with concrete and buildings all reduce the capacity of the soil to fulfill this need. As a result of an increased awareness of soil's importance, many changes are being made to protect soil. Recent interest in soil conservation holds the promise that humanity will take better care of this precious resource.

POOR SEEDLING EMERGENCE

Patchy crops reduce yield and profit. Poor seedling emergence is more of a problem under minimum tillage, because there is less soil disturbance and more stubble retention. It is less of a problem under conventional tillage, because the seedbeds are finely worked and the seeds are planted into bare ground. However, rather than reject minimum tillage because of its problems, it is preferable to address those problems.

POSSIBLE CAUSES

Poor seedling emergence may be due to:

- Poor seed–soil contact
- Inaccurate seed placement
- A soil temperature that is too low or too high
- Soil insects or soil-borne disease
- Surface crusting after sowing
- Poor quality seed.

Poor seedling vigour and establishment after emergence may caused by the presence of compacted soil beneath the seedling. The soil may have already compacted before sowing, or the planting tine may have smeared the bottom and sides of the seed trench.

Planting Machinery

If poor seedling emergence is due to poor seed–soil contact or inaccurate seed placement, then adjusting your sowing machinery correctly will help to improve emergence.

Surface Crusting

Rain after sowing and before seedling emergence may form a surface crust that reduces emergence. Once seedlings have emerged, the problem of surface crusting is less. However, until the crop reaches full ground cover (and protects the surface from raindrop impact) surface crusting will continue to reduce infiltration.

All soils seal under raindrop impact—even self-mulching soils seal if they are bare. On some soils, the seal is weak or it cracks as it dries; it offers little resistance to seedling emergence. However, on other soils, the seal forms a crust on drying, and this crust prevents seedling emergence. Further light rain will help emergence by softening the surface; further heavy rain will hinder emergence by strengthening the surface crust. Clay soils can crust if the surface is sodic. The treatment for such soils is gypsum.

COMPACTION MANAGEMENT

A small degree of compaction can be beneficial. Soil that is too loose may not be good for plant growth: a loose seedbed dries out too quickly and gives poor contact between the seeds and the soil; seeds may not germinate. The optimum level of compaction depends on the soil types, the plant, and the irrigation management during the growing season. Too much compaction can cause yield reduction. To reduce soil compaction:

- Time mechanical operations carefully
- Reduce axle loads
- Keep livestock off wet cropping country
- Confine traffic to laneways
- Use low ground pressure (wide) tyres (with caution).

The method of compaction repair will depend on the soil's capabilities. A cracking clay that swells and shrinks behaves quite differently to a hardsetting, fine, sandy or silty soil. If a soil swells and shrinks on wetting and drying you can use that natural action to break up a compacted layer. If a hardpan has developed in a soil with little clay, you will need to consider other options.

Example:

- Biological repair—that is, using growing plants to break up a compacted layer or increasing the soil organic matter content
- Mechanical repair—that is, deep ripping or mouldboard ploughing.

BIOLOGICAL REPAIR

Biological repair means using growing plants to restore soil structure. It usually involves a yield penalty. The plants will suffer in restoring the soil structure, and will produce a lower yield. Repair of compaction should therefore work in with the season, the market prices, and your long-term plans.

Biological repair includes:

- biological ripping (using plants to dry and crack the soil)
- Biological drilling (using tap-rooted plants to 'drill' through acompacted layer)
- Increasing soil organic matter content.

Biological Ripping

The easiest way to restore soil structure in compacted

cracking clays is to dry the soil, using any plants with a vigourous root system. Drying causes the soil to shrink and crack, allowing water, air and roots to penetrate through the cracks. The process is called biological ripping.

Biological Drilling of Non-Cracking Soils

In soils that don't crack, such as sandy, silty or loamy soils, 'biological ripping' is not an option. However, 'biological drilling' is possible. A strong tap-rooted plant such as lucerne can force its way through a compacted layer and leave root channels for subsequent plant roots to use. This is 'biological drilling'. The compacted layer must be moist: no root will enter dry soil. Depending upon how badly compacted the subsoil is, some roots will not be able to penetrate the compacted layer and the crop will suffer.

Increasing Soil Organic Matter

Some soils, such as fine sandy and silty soils, depend heavily on organic matter for the maintenance of their structures. Even cracking soils benefit from increased organic matter content. Increasing the organic matter of a soil content is a slow process. A pasture phase incorporated into your vegetable crop rotation is perhaps the most costeffective way to achieve this.

MECHANICAL REPAIR

Deep ripping and mouldboard ploughing are sometimes useful for quick results, but they are expensive operations. The initial success of the operation depends on the moisture content of the soil and the depth of tillage. Continuing success depends on managing traffic to avoid having machinery on the soil when it is too wet.

The Importance of Moisture Content

Soil moisture is important when you are working soil. A soil (particularly a clay) that is worked too wet will compact, while a silty or fine sandy soil that is worked too dry may powder. In both instances the soil will then structurally degrade. Before deep ripping, dig a hole to see if there is a compacted layer. If there is, check the layer's depth, thickness, and moisture content.

The severity of compaction will also influence your decision on deep ripping. If there is no compacted layer, there is no need to deep rip.

Deep ripping is expensive, and it could do more harm than good if the soil is too wet—a costly way of creating soil structural damage where none existed before! A compacted layer must be at the right moisture content to benefit from the ripping operation.

When you are cultivating clays (or when the compacted layer is clay) the layer must be dry enough to shatter, rather than smear. When the compacted layer is other than clay, some moisture is desirable so that the layer does not powder.

The depth and thickness of the compacted layer indicates how deep to set the ripper tine. To be effective, the tine should work just below the compacted layer. Dig a hole after a short run and check the effectiveness of the ripping operation.

Chapter 4

Diseases of Leaf

A number of leaf spotting or blotching diseases occur on trees and shrubs. Some of these are severe to plant health or aesthetics, such as rose black spot and scab on crabapple, but many are of minor effect. Following are descriptions of several of the more common leaf spot and leaf blotch diseases.

PLASMOPARA VIBURNI

This is one of the relatively few significant downy mildew diseases of ornamentals. Spores produced in dead plant tissue that overwinters splashes to foliage in the new season, causing patchy lesions on upper leaf surfaces and downy grayish-white fungal growth on lower leaf surfaces corresponding to these lesions.

Damage starts out with light greenish spots that grow together forming angular patches often bordered by veins. Leaf tissue in the blotches often reddens, browns, dries, shrivels and leaf drop sometimes occurs. The fungus reproduces rapidly and multiple infections occur during periods of leaf wetness and cool to warm, but not hot conditions.

Keep leaves as dry as possible by avoiding overhead irrigation and enhancing good air movement by proper plant siting and pruning practices. When applying preventive fungicides make sure to get good coverage of the lower leaf surfaces where infections occur. When fungicides are required, use a labeled product containing mancozeb. This disease is diagnostically different from powdery mildew of viburnum in that the fungus is present on lower rather than upper leaf surfaces.

BOTRYOSPHAERIA OBTUSA

This leafspot disease is a minor problem on most crabapples, although a few exhibit considerable leaf yellowing and leaf drop in some locations, including 'Madonna' and 'Professor Sprenger.' During moist spring weather the fungus infects leaf tissue and causes small roundish brown spots with purple borders. In some cases spots enlarge developing irregular lobes that, when they enclose the original roundish spots, cause a "frogeye" symptom.

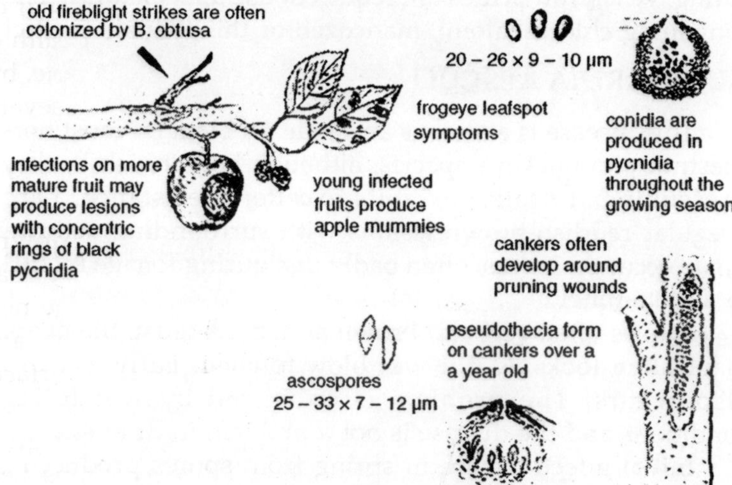

Fig. Life Cycle of Botryosphaeria obtusa

In addition to the leaf yellowing and defoliation that can occur on particularly susceptible cultivars, black rot cankers can develop on stems over the years, especially on crabapples weakened by winter injury and other environmental stresses. In the rare cases when fungicides are warranted, begin sprays at budbreak. When fungicides are used, apply labeled products containing mancozeb or chlorothalonil.

TAPHRINA CAERULESCENS

This is a minor disease in the northern states, occurring on both the red oak and white oak groups. Symptoms include light green to yellowish blister-like, roughly circular bulges on upper leaf surfaces (depressions as viewed from the lower leaf surface). These blisters tend to brown as the season progresses.

Spores overwinter on leaf buds and infect leaves as they open. This is the only infection that occurs each season, and leaves become resistant to infection as they mature. Mild, moist spring conditions favour infections. Controls are generally not warranted and fungicide recommendations are rarely made except in certain nursery/garden centre situations in which fungicides are applied dormantly before leaves emerge in the spring. When fungicides are required, use a labeled fungicide containing chlorothalonil, mancozeb or thiophanate-methyl.

GUIGNARDIA AESCULI

This disease is a serious aesthetic problem on most horse chestnut and buckeye species, although bottlebrush buckeye (Aesculus parvifolia) exhibits excellent resistance. Large irregular reddish-brown lesions with surrounding yellowed tissue occur on leaves, often badly disfiguring foliage by early to mid summer.

Leaves often curl and brown and, by August, the overall plant often looks as if it was blow-torched. Early leaf drop also occurs. The problem is enhanced by wet foliage conditions and the disease is not a problem in drier sites.

Initial infections are in spring from spores produced in infected leaves from the past year. Moist conditions enhance the infections and subsequent cycles of infection occur if moist conditions continue. Black fruiting bodies of the fungus are often evident in lesions.

The disease does not appear to be a serious health problem, as much of the annual growth of Aesculus has occurred by the time foliage is badly damaged. Controls for the serious aesthetic damage include fungicide applications made as leaves emerge, with repeated applications at 10 to 14 day intervals if wet conditions persist.

Use a labeled fungicide containing chlorothalonil or mancozeb. Also practice sanitation by cleaning up infested foliage at the end of the season, and improve air movement in the tree canopies to hasten leaf drying.

RHYTISMA ACERINUM, RHYTISMA PUNCTATUM

These dramatic but inconsequential diseases of numerous maple species cause small to almost one-inch diameter tar-like spots on leaves. The fungus overwinters on fallen leaves, then infects the upper surfaces of leaves in spring during moist conditions.

Leaf spots are first a yellowish green but by mid to late summer a tar-like mesh of fungal and leaf tissue develops inside the yellowed area. Occasionally some leaf withering and drop occurs but this is not generally serious and fungicide sprays are not generally recommended. If fungicides are required, use a labeled product containing mancozeb or triadimefon.

PHYLLOSTICTA MINIMA

Like many fungal leaf spots, this disease affects a number of maple species, most prominently Amur, Japanese, red and silver maple. The disease causes little damage because the infection is localized. Spots are roughly circular and develop into tannish spots with purple to red borders.

Later in the season the spots often contain black fruiting bodies of the fungus arranged in rings inside the lesion. Although this disease is quite noticeable in the landscape, especially on silver and red maples, and causes concern among homeowners, damage is minimal and fungicides are rarely necessary. If fungicides are required use a labeled product containing mancozeb or chlorothalonil.

Some common product names containing the fungicides mentioned in this fact sheet include: mancozeb (Fore, Dithane, Mancozeb); chlorothalonil; Thiophonate-methyl and triadimefon (Bayleton, Strike). It is the user's responsibility to be certain that the fungicide being used is labeled for the specific plant being treated.

CARE FOR TREE WOUNDS

One of the most common types of tree disease is wood or

trunk decay. Decay in a tree can result in dangerously weak trees, unsightly trees, or in shortening the life of the tree. Decay in a tree cannot be cured. Proper tree care can, however, prevent decay or greatly limit the progress of decay in the tree.

Decay is a condition that results from the digestion of wood by fungi and other microbes. Wounds through the bark start the processes that lead to decay. Wounds that expose the wood can be caused by animals, insects, birds, fire, storms, freezes, and the activities of humans, such as pruning. Of course, not all wounds lead to decay. Proper care of tree wounds can lessen the chance that decay will result.

When a tree is wounded, a natural process begins that results in the covering of the wound by bark and new wood. This is called wound closure, "callusing over," or "wound healing." Natural processes also begin that make wood beneath the wound unsuitable for growth of decay organisms. Proper treatment of wounds can hasten both of these processes.

Proper care of tree wounds begins with practices that promote callusing growth. Recent research work has shown that when a tree produces callus rapidly, the extent of the natural discolouration of the wood in the trunk above and below a wound is less than when callusing is poor. This is important because decay takes place only in discoloured wood. Rapid callus production commonly occurs in the spring when trees are growing vigorously.

Thus, to reduce decay hazard, the best time to make large pruning wounds is from the time that buds start to swell until the leaves are nearly fully expanded. This is usually in March, April, or May for much. The sort of large wounds we are referring to are those involving branches larger than 11/2 inches in diameter.

Fig. Trim Branches without Collars Flush to the Trunk.

In the past it has been recommended that the bark around ragged wounds on the trunk or large branches be trimmed away from the wound in an elliptical form, with the greatest dimension of the ellipse above and below the wound. Recent research suggests that when wounds are made in the spring (as recommended above) this is not a good practice.

This is because trimming away healthy bark may lead to increased wood discolouration and hence to increased decay. Elliptical trimming may not be important when wounds occur in late summer or fall either. The bark over these sorts of wounds tends to die and become loose in an elliptical fashion naturally.

Healing associated with branch stubs may often be delayed and lead to increased decay within the tree if the stub is not properly pruned off. Cutting flush with the tree is generally best. Some branches have collars, "shoulders" or other natural projections.

These should not be trimmed off. When natural projections or collars are removed, not only is there a larger area to be callused over, but chances of decay organisms entering the wound and becoming established are greater. Branch stubs previously incorrectly pruned should be corrected as promptly as possible. Care should be taken, however, not to cut into the new callus tissue that has begun to form in response to the wound.

Try to keep wounded trees growing vigorously. This will increase the rate of wound closure and reduce the amount of discoloured wood and decay associated with wounds. Trees should be fertilized properly and watered during droughts. Aerate the soil if it is compacted. Do not plant trees that will grow to be large too close to a house, paved area, or other restrictive object or soil situation.

Thin out less valuable, competing trees. Take measures to control debilitating insect infestations. Protect the tree from wounds made by careless mowing or weeding practices. Prune badly broken branches promptly. You may use wound dressings if you wish; however, some tree care specialists feel that wound dressings interfere with healing or callusing. A light application of tree wound dressing does indicate that someone attended to the problem and may make a pruning job look better or more professional.

Trees in the urban landscape must be helped. Early in the life of the tree, establish sound maintenance programmes for all trees. Part of such maintenance is the prevention of wounds, maintenance of tree vigour, and proper wound care.

SOOTY MOLDS ON TREES AND SHRUBS

Symptoms

A sooty, gray-black, velvety, often crust-like coating may develop on the leaves or needles, fruits, and branches of certain plants. The coating is actually the growth of one of several species of black-coloured fungi or molds. The coating can be removed easily by rubbing the leaf between the fingers, thus exposing the green leaf tissue below. Sooty molds grow only on the plant surface and will not kill plants. In fact, sooty molds often grow on sidewalks or fences under infested trees. Sooty molds are normally considered to be a cosmetic or aesthetic problem.

In extremely severe cases, it is possible for the black growth to block enough sunlight to interfere with photosynthesis. In such cases, leaves or needles, fruits and new shoots may be smaller, or less intensely coloured. Respiration can be reduced through the physical closure of stomates by the molds' vegetative growth. Under drought conditions, plants affected with sooty mold will wilt more rapidly than unaffected plants. If plant vigour has been reduced, the plant may also be predisposed to further injury by other insects, diseases or environmental stresses.

Insect Association

Sucking insects are the primary cause of sooty mold growth. Many plant sap-sucking insects feed on leaves and stems of trees and shrubs. These sucking insects often produce excessive, watery excrement rich in sugars. This excrement is called honeydew. Excreted honeydew often falls on leaves or needles, branches, fruits or anything else immediately underneath the infested area of the plant. It is on this excretion that the sooty mold fungi grow. Sometimes plants not actually infested by insects may be affected if a tree above them is being attacked by a honeydew producing insect and the honeydew drops onto them.The major types of honeydew producing insects are: Aphids, soft scales, mealybugs, whiteflies, leafhoppers, planthoppers and psyllids. Occasionally, the spittle-like froth produced by spittlebugs promotes the growth of sooty molds.

Susceptible and Resistant Trees

Plants that are commonly infested by the insects mentioned above are often hosts to sooty molds. Resistance to this condition is normally a result of resistance by the plant to honeydew producing insects.

Control

Control sooty molds by controlling the honeydew producing insect. Chemical control is most commonly used to manage these sucking insects, although biological and cultural control strategies are available for some aphids and scales.One must identify the insect that is causing the problem and the plant being attacked before deciding on a method of management. Consult an appropriate up-to-date OSU Extension fact sheet or bulletin such as Bulletin 504 (Insect and Mite Control in Woody Ornamental and Herbaceous Perennials). A strong spray of water can be used to dislodge the mold growth from many plants.

POWDERY MILDEWS ON ORNAMENTAL PLANTS

Almost all landscapes have plants that become diseased with one of the powdery mildew fungi. Although the fungi that cause powdery mildew are usually different on different plants, all of the powdery mildew diseases are similar in appearance. In most cases, prompt recognition and control actions can prevent severe damage to plants from powdery mildew diseases.

Symptoms

Powdery mildews, as the name implies, often appear as a

superficial white or gray powdery growth of fungus over the surface of leaves, stems, flowers, or fruit of affected plants. These patches may enlarge until they cover the entire leaf on one or both sides. Young foliage and shoots may be particularly susceptible.

Leaf curling and twisting may be noted before the fungus is noticed. Severe powdery mildew infection will result in yellowed leaves, dried and brown leaves, and disfigured shoots and flowers.

Although it usually is not a fatal disease, powdery mildew may hasten plant defoliation and fall dormancy, and the infected plant may become extremely unsightly. On roses, uncontrolled powdery mildew will prevent normal flowering on highly susceptible cultivars.

Hosts

Powdery mildew fungi infect almost all ornamental plants. They are commonly seen only on those plants more naturally susceptible to the disease. Susceptible woody plants include some deciduous azaleas, buckeye, catalpa, cherry, a few of the flowering crabapples, dogwood, English oaks, euonymus, honeysuckle, horse chestnut, lilac, privet, roses, serviceberry, silver maple, sycamore, tulip tree, some viburnums, walnut, willow and wintercreeper.

Powdery mildews are also common on certain herbaceous plants, such as chrysanthemums, dahlias, delphiniums, kalanchoes, phlox, Reiger begonias, snapdragons and zinnias.Remember that each species of powdery mildew has a very limited host range. Infection of one plant type does not necessarily mean that others are threatened.

Environment Favouring Powdery Mildews

Most powdery mildew fungi produce airborne spores and infect plants when temperatures are moderate (60 to 80 degrees F) and will not be present during the hottest days of the summer. Unlike most other fungi that infect plants, powdery mildew fungi do not require free water on the plant surface in order to germinate and infect. Some powdery mildew fungi, especially those on rose, apple, and cherry are favoured by high humidities. Overcrowding and shading will keep plants cool and promote higher humidity. These conditions are highly conducive to powdery mildew development.

Control of Powdery Mildews

Before using fungicides you should attempt to limit powdery mildews by other means. The following cultural practices should be beneficial for controlling powdery mildews.

- Purchase only top-quality, disease-free plants of resistant cultivars and species from a reputable nursery, greenhouse or garden centre. Horticulturists in the green industry and Extension offices should be consulted concerning the availability and performance of resistant varieties.
- Prune out diseased terminals of woody plants, such as rose and crabapple, during the normal pruning period. All dead wood should be removed and destroyed (preferably by burning). Rake up and destroy all dead leaves that might harbor the fungus.
- Maintain plants in a high vigour.
  - Plant properly in well-prepared and well-drained soil where the plants will obtain all-day sun (or a minimum of 6 hours of sunlight daily).
  - Space plants for good air circulation. DO NOT plant highly susceptible plants—such as phlox, rose, and zinnia—in damp, shady locations
  - Do not handle or work among the plants when the foliage is wet.
  - Water thoroughly at weekly intervals during periods of drought. The soil should be moist-8 to 12 inches deep. Avoid overhead watering and sprinkling the foliage, especially in late afternoon or evening. Use a soil soaker hose or root feeder so the foliage is not wetted.

Chemical Control of Powdery Mildews

In many cases, powdery mildew diseases do little damage

to overall plant health, and yearly infections can be ignored if unsightliness is not a major concern. For example, lilacs can have powdery mildew each year, with little or no apparent effect on plant health. On some plants, powdery mildews can result in significant damage. Thus, fungicides must be used to achieve acceptable control.

ANTHRACNOSE LEAF BLIGHT OF SHADE TREES

A number of different trees are affected by anthracnose diseases. These fungal diseases can cause severe leaf blighting and deformation, but in many cases damage to plant health is not severe. However, with sycamore anthracnose and dogwood anthracnose, the fungus regularly moves back into stem tissue and causes more significant problems. Following are profiles of some of the more common anthracnose diseases of landscape trees.

KABATIELLA APOCRYPTA

Anthracnose diseases are generally not severe on maple, but can cause considerable unsightliness from brownish leaf blotches and some leaf drop when moist weather conditions make the disease particularly severe. Extensive development of stem infections is not common on maples, as it is with sycamore anthracnose and dogwood anthracnose.

The most common symptoms include brownish discolouration along veins, varying from discrete spots to irregular patches of discolouration bordered by veins. Spore masses of the fungus can sometimes be found on lower leaf surfaces along veins during extended moist conditions. The fungus spreads from previously infected tissue in spring to new growth. Where fungicides are used, applications must be started at bud break and continued during early leaf development. When fungicides are required, use labeled products containing mancozeb, thiophanate-methyl, or chlorothalonil.

APIOGNOMONIA ERRABUNDA

This anthracnose disease is primarily a leaf blighting and

blotching disease of white ash and, to a lesser extent, green ash. Small twig cankers do occur and the fungus overwinters on these twig lesions, but little damage occurs from this phase of the disease. In wet, cool spring conditions, leaves and sometimes shoots first develop water soaked areas and later large tannish blotches and leaflet distortion.

Considerable leaf drop occurs, especially from lower areas of the canopy. Though this causes concern when leaves litter the ground in late spring, damage to overall plant health is not generally severe and plants typically releaf. As leaves mature they tend to become more resistant to infection. Fungicide applications, if warranted, should be made at bud break, with several repeat applications early in the season. When fungicide is required, use a labeled material containing thiophanate-methyl, chlorothalonil, or mancozeb.

APIOGNOMONIA QUERCINA

White oaks are the most susceptible of many oak species to this leaf blotching disease. Twig infections occur but are not significant except as sources of overwintering fungal inoculum from year to year. Leaves and shoots are infected during cool, wet spring conditions causing leaf blotches that often are strictly delimited by leaf veins.

Eventually, lesions become a papery tan colour and some leaf shriveling occurs. Multiple cycles of infection can occur. Just as leaves near maturity, lesion size lessens, and once leaves mature they become fully resistant by early to mid summer. Fungicides are generally not recommended. If fungicide is required, use a labeled material containing thiophanate-methyl, chlorothalonil or mancozeb.

APIOGNOMONIA VENETA

This is a potentially serious disease of American sycamore and to a lesser extent London planetree. Susceptibility of London planetree varies considerably with seed source. Shoot blight, leaf blight and twig and branch cankers and dieback can be severe. The fungus overwinters on twig tissue on the tree with spores splashing to new buds, shoots and leaves in the spring, with disease being enhanced by cool, wet conditions during shoot and leaf development. Considerable defoliation, sometimes with complete leaf loss, occurs on many trees by late spring in some years.

Trees typically releaf by early to mid summer and are less susceptible to continued infections because of warmer, drier conditions. Also, as leaves age they become less susceptible to infection. With repeated infections over the years, cankering of twigs and branches can result in erratic shoot growth that gives an overall distorted appearance to the tree, and also witches-brooming where there are numerous side shoots that develop around a central terminal shoot that was killed by the fungus.

Fungicide applications to prevent infections in the spring are sometimes warranted. If fungicide is required, use a labeled material containing thiophanate-methyl, chlorothalonil or mancozeb. Fungicide injections (which are made by tree care professionals) are also used in spring and fall to systemically control the disease.

DISCULA DESTRUCTIVA

In recent years this disease has become prominent in certain areas, especially on flowering dogwood (Cornus florida). Dogwood anthracnose is most severe where cool, moist conditions occur during the summer, such as in higher elevation areas, and in densely vegetated shady sites with poor air movement. Leaf symptoms include irregular brown blotches bordered in purple on upper leaf surfaces (tan in colour viewed from the leaf underside). Leaf lesions often are delimited by the leaf midvein.

Stem symptoms include twig dieback and stem cankering and dieback, often with visible fungal fruiting bodies on dead twigs. Attached wilted, brown leaves often persist into the next spring instead of dropping in the fall.

Plants may be killed. Fungal infections occur during moist conditions and fungicides are recommended in the spring from the period of early bud break through bract fall and through early leaf development. When fungicide is required, use a labeled product containing propiconazole or mancozeb. Flowering dogwood should be planted in sites with good soil drainage and adequate organic matter.

Partially shaded sites, such as those with just morning sun are ideal. Plants should receive a moderate fertility programme, and should be mulched to moderate fluctuations in soil temperatures. Dead branches should be promptly pruned from the plant.

Control of Anthracnose Diseases

- Overall tree care programme. Use proper fertilization, pruning, watering, and pest control practices to encourage vigorous plant growth. This aids in general tolerance of the effects of disease and in rapid refoliation in years where disease is severe.
- Fungicide applications. If significant damage occurs yearly and controls are justified, properly applied fungicides may reduce damage from these diseases. High pressure spray equipment will be needed for large tree applications, and this typically requires the hiring of a professional tree care service.

Fungicides will not be effective unless they are applied before and during infection periods. Typically, three applications are necessary, beginning in early spring, with the first application made before leaf buds open. Applications in the fall have been shown to be useful for sycamore anthracnose control. Fungicide injections have also shown promise for sycamore anthracnose control. These must be applied by professional tree care companies.

VERTICILLIUM WILT OF LANDSCAPE TREES AND SHRUBS

Verticillium wilt, caused by the fungi Verticillium albo-atrum and V. dahliae, is a serious vascular disease of hundreds of woody and herbaceous plant hosts. Food crop hosts include everything from raspberries and strawberries, to tomatoes and potatoes.

Some of the many common woody ornamental host plants

include ash, barberry, catalpa, elm, magnolia, maple, Russian olive, redbud, smoketree, tuliptree, and viburnum. One group of plants not susceptible to Verticillium wilt are all the gymnosperms, including conifers such as pine and spruce. While many landscape plants are affected, Verticillium wilt is not a major problem in natural forested areas.

Diagnostic Symptoms

Wilting of leaves and dieback of branches, often one at a time or on one side of the tree, are the most severe symptoms. This can occur over a number of years, with remission of symptoms in some years, or can rapidly progress to plant death in a year or two. Other symptoms of Verticillium wilt may include: marginal browning and scorch of leaves, abnormally large seed crops, small leaves, stunting, poor annual growth, and sparse foliage.

Sometimes large areas of cambial tissue die from infections by the fungus and opportunistic fungi such as Nectria develop in elongated cankers. Late season infections may not be noticeable until plants come out of dormancy with branch dieback evident.

All of the above symptoms can also be caused by other stress factors. A good field symptom that can set Verticillium wilt apart diagnostically is the discolouration of xylem and cambial tissue, visible as streaks if you cut into the wood. This discolouration is variable for different plants: generally greenish to blackish on maple, yellowish green on smoketree, and brown on ash.

This streaking is not totally diagnostic on two counts:

- Other fungi and other factors can cause discolouration; and
- On some hosts and on youngest twigs, infection is not always accompanied by discolouration.

Disease Cycle and Conditions Favouring Disease

The Verticillium fungus can survive for many years in soil, making effective crop rotation difficult. The fungus infects plant roots through wounds and in some cases direct penetration of root tissue. Verticillium also is transmitted from plant to plant by grafting and budding. From root infections, the fungus spreads upward in the plant through the vascular stream.

The results of infection are tissue damage and plugging of xylem, robbing stems and leaves of needed water and minerals. The fungus is returned to the soil as plant parts fall or die, and tiny resistant fungal microsclerotia are spread by wind, in soil and on equipment. Many weed hosts are also susceptible; therefore the cycle of contaminated soil is hard to break. Development of Verticillium wilt is favoured by factors that stress roots, including wounding and droughty conditions.

Control

Disease resistance. If Verticillium wilt is diagnosed at a particular landscape or nursery site it is prudent to replant into that area with a plant that exhibits resistance to this disease. A few common examples of plants typically free of this disease include: crabapple, mountain ash, beech, birch, boxwood, dogwood, sweet gum, hawthorn, holly, katsuratree, honeylocust, oak, pear, London planetree and sycamore, rhododendron, willow, and zelkova. The red maple cultivars Armstrong, Autumn Flame, Bowhall, October Glory, Red Sunset, Scarlet and Schlessinger have also been reported as resistant.

Keep plants as healthy as possible. Proper transplanting practices, proper water management to avoid droughts, a good fertility programme, and pruning out dead branches are all good plant health care management practices. These can help limit infections and help limit the effects of these infections to some extent. Pruning out infected branches is useful as a general horticultural practice for overall plant vigour and aesthetics, but does not eliminate Verticillium from the plant since infections spread from the roots.

Fungicides are not effective for control of this disease.
Bacterial Crown Gall on Ornamentals in the Landscape
Crown gall is characterized by growth of galls on roots or

stems. While mostly found on woody plants, it affects some herbaceous plants as well. Although found on more than 600 plant species in over 90 families, the disease is of economic importance on relatively few ornamental plants. Some commonly affected ornamentals include rose, Prunus, (flowering cherry, flowering almond and ornamental plums), willow, and certain Euonymous species, especially wintercreeper.

Effects of Crown Gall

Crown gall can reduce the productive life of plants. Deformation of tissues due to gall formation disrupts the movement of water and nutrients between roots and leaves. Stems are weakened and growth may be reduced with a general decline in vigour. The severity of the disease depends on the size, number, and location of the galls, and also on the susceptibility of the plant and age when infected. Galls at the crown of young plants have the greatest adverse effect and can cause stunting and failure to produce healthy leaves and blossoms. This disease may have little noticeable effect on older plants.

Symptoms

Galls may develop anywhere on stems and roots, but are usually found near the soil line. They vary from pea size up to several inches in diameter. Young galls are light coloured and smooth. Older galls become discoloured, hard and woody, and eventually crack, decay and slough off. The texture at first is softer than the normal host stem or root tissue. The galls consist of disorganized host tissues. Secondary galls sometimes form above the sites of the primary gall on stems of some hosts. The secondary galls are usually smaller and occur as separate or unbroken elongated masses of tissue breaking through the bark surfaces. Unlike insect galls, crown galls are a solid mass of tissue all the way through.

Disease Development

Crown gall is caused by the soil-borne bacterium,

Agrobacterium tumefaciens. The bacteria can persist in the soil for two or more years even in the absence of susceptible plants. Sometimes the bacteria are carried on seeds. Fresh wounds in stems or roots are essential for the bacteria to invade host tissues. These wounds commonly occur during planting, cultivation and pruning, and during propagation when grafting and taking cuttings. Soil insects and nematodes can also cause root wounds providing entry sites.

There are no effective chemical controls for this disease in the landscape. Cultural controls include:

- Avoid unnecessary wounding (protect from injury).
- Sanitation-remove infected plants.
- Plant resistant plants in crown gall-infested areas.
- Do not purchase plants with suspicious swelling near the soil line or on the roots.

CEDAR RUST DISEASES OF ORNAMENTAL PLANTS

There are a number of "cedar rust" diseases in which the fungus completes its life cycle on two plant hosts; one in the cypress family and one in the rose family (the rosaceous host). Discussed here are three common cedar rust diseases in the northeast U.S.

- Cedar apple rust (pathogen: Gymnosporangium juniperi-virginianae). The fungus alternates between Eastern red cedar (Juniperus virginiana) and mostly apple and crabapple.
- Cedar hawthorn rust (pathogen: Gymnosporangium globosum). The fungus alternates between junipers and hawthorn, crabapple, and apple in addition to several other rosaceous hosts.
- Cedar quince rust (pathogen: Gymnosporangium clavipes). The fungus alternates between junipers and a wide range of rosaceous hosts. The most noticeable in the landscape is hawthorn.

In some cases these diseases are minor problems, but cedar quince rust and cedar hawthorn rust can be a major problem on hawthorns and cedar apple rust is a major economic consideration in commercial apple production.

Diagnostic Symptoms

Cedar apple rust: On junipers, tan to brownish round to kidney-shaped fungal galls are present in winter and early spring. With moist weather, gaudy bright orange masses of gelatinous spores develop from these galls, and galls swell to several times their original size. Spore masses are several inches in diameter, with a central core and radiating hornlike tendrils, and are highly visible during moist weather in mid-spring.

On apple and crabapple, bright orange-yellow leaf spots develop on upper surfaces of leaves in late spring, followed by light coloured, fringed cup-shaped structures on lower leaf surfaces several weeks later. Damage on junipers is generally minor and involves presence of the galls and twig dieback. On apples and crabapples, fruit infections and leaf drop also can occur.

Cedar hawthorn rust: On junipers, galls are somewhat smaller than with cedar apple rust disease. Galls continue to produce spores on junipers for more than one year, compared to only one season of spore production with cedar apple rust. On hawthorn, leaf spots are similar to above and occasionally green twigs are deformed by the fungus.

Cedar quince rust: Infected areas on juniper are much less spectacular than with cedar apple rust, with a cushion-like mat of orangish fungal growth developing on spherical galls in spring. Cedar quince rust causes the greatest damage of the three rusts to ornamental rosaceous hosts, especially to hawthorns, because of extensive, unsightly fruit infestations, stunting and death of fruits and swelling and distortion of twigs. Infected leaves brown and die. Fruits become covered with orangish-pink spore horns. Unsightly spherical cankers developing on stems can last more than one year.

Disease Cycle and Conditions Favouring Disease

Rust fungi have complicated disease cycles with a number of different spore types that will not be detailed here. A crucial factor relative to control on these cedar rusts, however, is that there is no repeating spore cycle on the rosaceous hosts. In other words, spores produced on hawthorn will not reinfect hawthorns or other rosaceous plants-they will only reinfect junipers later in the season. Spores produced on juniper will not reinfect junipers-they will only infect the rosaceous host. The alternating host plant is necessary for survival of the fungus.

Spores produced on the juniper host are blown during moist weather to the rosaceous hosts in mid-spring at a time when new growth has emerged. The fungus then causes leaf spots on upper leaf surfaces and while growing in the leaf two strains of the fungus mate and emerge as a new spore form on the lower leaf surface.

These spores are then blown back to junipers in mid summer to fall, develop galled areas on the junipers over a one and a half year period and the cycle begins again. Windborne spread of spores between the hosts of several hundred yards is not unusual and spread can be a matter of miles.

Control

Application of fungicides. Protective fungicides can be applied several times starting with prebloom on hawthorn and bud break on crabapples if the disease is chronically a problem at a given site. These applications are to protect the plant from spores being disseminated from the juniper host in mid-spring. Since there is no repeating cycle of this disease on the rosaceous host, further applications after this springtime spread from juniper are unnecessary.

Commonly recommended fungicides include: Mancozeb (Fore, Dithane, Mancozeb); Chlorothalonil (Daconil*); Triadimefon (Bayleton, Strike) and propiconazole (Banner). It is the user's responsibility to follow all label instructions.

When you diagnose cedar rust disease from infected hawthorn or crabapple fruits and leaves it is far too late to spray for that year. Sprays are rarely recommended to protect the juniper host from spores being disseminated from the rosaceous host in late summer and fall.

Eradication of the other host plant. One approach

sometimes suggested is to eliminate junipers from around plantings of rosaceous hosts, and vice versa. Concerted efforts to eradicate junipers were historically tried in concentrated apple growing regions.

This practice is limited to some extent by practicality in terms of the widespread occurrence of junipers, long distance spread of the fungi involved, the rights of juniper lovers, and the fact that in most situations cedar diseases are not so serious that such extreme measures are needed. Nevertheless, it is prudent to separate highly susceptible junipers and rosaceous hosts to the extent possible in nursery and landscape situations.

One simple practice where only a few plants are involved is to remove galls from junipers. This is easier to do with cedar apple rust and cedar hawthorn rust, since galled areas are more inconspicuous with cedar quince rust.Use plants with genetic resistance.

A number of juniper species and cultivars and a number of rosaceous plant species and cultivars have varying levels of resistance and susceptibility to these three diseases and where disease pressures are historically high these plants should be used.

TIP BLIGHTS OF JUNIPERS

Phomopsis tip blight and Kabatina tip blight are two common diseases of junipers found in most states east of the Mississippi. Both diseases are caused by fungi and the damage they cause on nursery stock, transplants and certain juniper varieties in the landscape can be severe; however, most established junipers in the landscape are seldom killed. The disease is most serious on younger plants and becomes less serious as plants get older.

There are many varieties of juniper that vary from very susceptible to highly resistant. Junipers are generally considered as low maintenance because they are relatively free of major diseases and insect pests; however, these diseases can adversely affect the appearance and health of these trees in certain locations and under the proper environmental conditions. Although Phomopsis and Kabatina blights cause almost identical symptoms, aspects of their development and control do differ. Therefore, it is important to distinguish between the two diseases.

Symptoms and Causal Organisms

Phomopsis tip blight, caused by the fungus Phomopsis juniperovora, damages new growth and succulent branch tips of junipers from mid-April through September. Older, mature foliage is resistant to infection; therefore, most blighting occurs on the terminal 4 to 6 inches of the branches.

Affected foliage first turns dull red or brown and finally ash-gray. Small gray lesions often girdle branch tips and cause blighting of foliage beyond the diseased tissue. Small, black, spore-containing fungal fruiting bodies develop in the lesions. Use a hand lens to view these diagnostic fungal structures more easily.

Spores of the Phomopsis fungus are produced throughout the summer, and infection can occur whenever young foliage is available and moisture or humidity are high. Most infections usually occur in April through early June and again in late August through September. Very few infections occur in mid-summer or during the winter months.

Repeated blighting in early summer can result in abnormal bunching (witches' broom) and discolouration of the foliage, stunting of young trees or shrubs, or-in severe cases-plant death. Be cautious in diagnosing witches brooming and stunting because similar damage can be caused by the dwarf tip mite. With juniper problems, it is always a good idea to have problem diagnosis confirmed by a diagnostic laboratory.

Kabatina tip blight, caused by the fungus Kabatina juniperi, first appears in February and March; and well before symptoms of Phomopsis tip blight appear. The terminal 2 to 6 inches of diseased branches throughout the juniper first turn dull green, then red or yellow.

Small ash-gray to silver lesions dotted with small, black fruiting bodies of the fungus are visible at the base of the discoloured tissue. The brown, desiccated foliage eventually drops from the tree in late May or June. Foliar blighting occurs only in early spring; it does not continue through the summer. Blighting is also restricted to the branch tips and does not cause extensive branch dieback or tree death. Be cautious in diagnosing witches brooming and stunting symptoms because similar damage can be caused by the dwarf tip mite. With juniper problems it is always a good idea to have problem diagnosis confirmed by a diagnostic laboratory.

The primary infection period for the Kabatina fungus is thought to be in autumn even though visible symptoms are not apparent until late winter or early spring. Infection often is associated with small wounds on branch tips caused by insect feeding or mechanical damage.

Control

- When purchasing new plants, select those that have been reported to have disease resistance. Table has information on relative disease resistance to Phomopsis and Kabatina for several Juniper selections.
- Space new plantings to provide good ventilation and air circulation and avoid heavily shaded areas. Avoid wounding plants, especially in spring and autumn.
- Water plants in early morning so the foliage will dry as soon as possible. Maintain adequate fertility, but do not over fertilize.
- Prune out diseased branch tips during dry summer weather and destroy them. Do so only when plants are dry and no rain or overhead irrigation is expected for several days. Avoid excessive pruning or shearing.
- Chemical control of these tip blight diseases normally is not necessary in established landscape or windbreak plantings. Occasionally, fungicide applications may be needed on susceptible junipers to control Phomopsis blight. Application of certain copper-based fungicides (Phyton-27, Kocide), thiophanate-methyl (Cleary's 3336, Domain, Fungo FLO), or mancozeb (Fore, Dithane, mancozeb) at 7- to 21-day intervals during rapid plant growth in the

spring will give adequate control of Phomopsis but not Kabatina tip blight. Kabatina blight infections occur in the fall, and there currently are no fungicides labeled for control of this disease.

RHIZOSPHAERA NEEDLECAST ON SPRUCE

Spruce, in particular, Colorado blue spruce, can be infected with a needlecast disease caused by the fungus *Rhizosphaera kalkhoffii.* Trees planted in nurseries, Christmas tree plantations, and landscapes can be infected. Trees are not usually killed by this disease; however, premature needlecast results in trees that are not marketable, or which are not acceptable in the landscape.

Symptoms and Disease Cycle

A healthy spruce will retain its needles 5 to 7 years. A spruce severely infected with Rhizosphaera needlecast may hold only the current year's needles. Rhizosphaera needlecast infects needles on the lower branches first and gradually progresses up the tree. This pattern holds true for most needle diseases on conifers and is the result of more favourable conditions for disease development near the ground. Under epidemic conditions, lower branches may be killed by this fungus. Although needles on new growth become infected in May and June, symptoms are not visible until late fall or the following spring, when infected needles turn purple to brown and begin to drop. Tiny fruiting bodies of the Rhizosphaera fungus protrude through the stomata of the infected needles. Under a hand lens, these stomata appear as fuzzy black spots instead of their usual healthy white colour. During wet weather in late spring, spores are released from these fruiting bodies and are rain splashed onto newly developing needles where infection occurs and the disease cycle is repeated.

Cultural Control

Very little is known about cultural control of Rhizosphaera needlecast. The following guidelines will help prevent serious losses.

Use Healthy Stock

At planting time, the foliage of blue spruce should be examined for fruiting bodies of Rhizosphaera protruding through needle stomata. If these bodies are present, the tree should not be planted.

Maintain Tree Vigour

Although detailed studies are lacking, it has been observed that trees suffering from environmental stresses are often more seriously attacked by Rhizosphaera. Spruce are particularly sensitive to heavy, compacted soils which become quite dry in late summer. Vertical mulching such soils to improve aeration and water penetration may help lessen the severity of the disease. Root irrigation during dry weather should also be carried out whenever possible.

Prevent Spread by Shearing Tools

Shearing when the foliage is wet may result in spread of the spores on shearing tools. To avoid this possibility, do not shear infected trees when the foliage is wet (such as when dew is on the foliage in the morning).

Shear healthy trees first to avoid carrying the spores from a diseased tree to a healthy one. If this is not possible, tools should be sterilized after shearing a diseased planting. Denatured alcohol, available at most paint stores, will kill the spores and also remove pitch from tools. A three- to five-minute dip will do the job.

Chapter 5

Diseases of Corns

The major corn diseases can be grouped into four categories: leaf blights, stalk rots, ear rots, and viral diseases.

LEAF BLIGHTS

A number of leaf-blight diseases occur on corn. The most common are gray leaf spot, Stewart's bacterial leaf blight, and northern corn leaf blight. These diseases can be found in almost any field, depending on the year and susceptibility of the hybrid planted. Some leaf-blight diseases are most often found associated with continuous corn, especially in reduced-tillage, continuous corn fields. These are anthracnose, gray leaf spot, eyespot, and northern leaf spot.

All leaf blight diseases cause loss of green leaf tissue, resulting in fewer kernels and lightweight grain. Plants may be predisposed to stalk-rot diseases when leaf damage is severe. The amount of yield loss is usually related to the time when the plant's upper leaves become infected.

The most severe yield loss occurs when the upper leaves, the ear leaf, and those above the ear, become infected at or soon after tasseling. Yield losses will be minimal if disease does not occur on these leaves until six to eight weeks after tasseling.

Leaf blight diseases are most effectively controlled by selecting hybrids with genetic resistance. Contact your seed dealer for information on hybrids with resistance to gray leaf spot, Stewart's bacterial leaf blight and other leaf diseases important in your area. A one-to two-year rotation away from corn and destruction of old corn residues by tillage may be helpful if susceptible hybrids must be grown. Fungicides are also available for control of leaf diseases, but are economically viable only under severe disease pressure.

Stalk Rot

Stalk rots are the most important and common diseases of corn. Annual losses are estimated at 5 to 10 per cent. There are several stalk-rot diseases, but Gibberella stalk rot and anthracnose stalk rot currently are the most prevalent. Both are fungal diseases that result in premature ripening, chaffy ears, and lodging of plants before harvest. The interior of the stalk becomes rotted, tissues break down, and the stalk is easily broken.

Anthracnose stalk rot is usually associated with continuous corn and is recognized by the blackening of the outer surface of the stalk late in the season.

Stalks with Gibberella stalk rot can be found in nearly any field. Affected stalks often have pink to reddish discoloured internal tissues. Control of stalk rot diseases is based on reducing plant stress from factors such as lack of moisture, leaf diseases, insect injury, and nutritional stress. To reduce the affects of stalk rot diseases, follow as many of the following practices as possible:

- Select hybrids with good standability and resistance to leaf blight diseases.
- Adjust soil fertility to recommendations based on a soil test. Avoid excessive rates of nitrogen in relation to potassium.
- Follow a one- to three-year rotation away from corn. Soybeans, forage legumes, and small grains are acceptable in the rotation. The longer the rotation away from corn the better.
- Plant at populations recommended for the hybrid grown. Overplanting leads to increased moisture, light and nutrient competition, and more plant stress.
- Harvest fields with the greatest level of rotted stalks first to avoid lost ears on lodged plants.
- Control insects, particularly root worms and stalk

borer. Insects cause injuries to plant roots and stalks permitting stalk rot fungi to enter the plant.

Ear Rot

Gibberella, Fusarium, and Diplodia ear rot diseases, but Gibberella ear rot is the most important. The Gibberella ear rot fungus is the same fungus that causes Gibberella stalk-rot disease. *Gibberella* enters from the silk end of the ear when cool, wet weather persists for several weeks through late silking of the crop. The occurrence of a whitish to pinkish mold on the ear tip is diagnostic, but extensive mold growth may not occur. On shelled grain, the symptoms may be seen as a pinkish colouration in some of the kernels.

Even though extensive rotting does not always occur, the disease is serious because the fungus frequently produces toxins that makes the corn unfit for feeding. Hogs are particularly sensitive to the toxins produced in moldy grain and may refuse to eat it even when hungry. If hogs refuse to eat grain, have a mycotoxin analysis run to determine the kinds and levels of toxin present. Some corn hybrids are less susceptible than others to Gibberella ear rot.

Ears with tight husks which mature in an upright position often have more ear rot than those maturing in a declined position. Diplodia ear rot appears to be more common in continuous corn under reduced tillage. Ears affected by *Diplodia* are covered with a thick mat of white fungal growth. Fusarium ear rot is common, but only individual kernels are affected on ears. Plant hybrids known to be less susceptible to these ear diseases.

Grain with evidence of ear or kernel rot should be dried to 14 per cent moisture before storage and maintained at this level until used. Feeding less than 5 per cent moldy kernels may prevent feeding problems, but grain should be tested for mycotoxin levels prior to feeding to avoid any problems.

Virus Diseases

Maize dwarf mosaic and maize chlorotic dwarf, are potentially destructive diseases where johnsongrass is established. The two viruses that cause these diseases are able to survive in this perennial weed grass. Aphids and leafhoppers feeding on johnsongrass in the spring pick up the virus and inoculate nearby corn. Control is achieved by planting resistant or tolerant hybrids. Efforts also should be made to eradicate johnsongrass.

OAT DISEASES

In the past, oat diseases have resulted in considerable yield losses. Many of these diseases are no longer important since the development of resistant varieties. Diseases that have caused problems in recent years include loose smut, covered smut, and barley yellow dwarf virus. Several new resistant varieties are available for control of barley yellow dwarf virus. diseases should be planted.

ALFALFA DISEASES

There are more than a dozen diseases of alfalfa. They can be grouped into two categories: those that affect the stems, crowns, or roots and those affecting the foliage. The stem and root-rot diseases are the most serious. Major losses of stand have been caused by Phytophthora root rot, anthracnose, and Sclerotinia crown rot. Verticillium wilt is a constant threat because some farmers are still growing old varieties that lack resistance to this disease.

Phytophthora root rot has been responsible for loss of stand in the seedling year when rainfall is above average and/or surface and subsurface drainage is poor. It is caused by a soilborne fungus that becomes active under wet soil conditions and may attack both seedlings and older plants. Dark brown, decayed areas on the tap root two to three inches below the soil surface are early indicators of Phytophthora root rot. Soil moisture is the key factor affecting disease development.

Any improvement in surface and subsurface drainage will reduce losses from this disease as well as from other less common root rots. Varieties with good levels of resistance are now available. *Phytophthora* specific fungicide seed treatments are available to help prevent loss of stands due to seedling damping-off. Fungicides are labeled for use at planting either as a broadcast application or incorporated on fertilizer granules and applied in a band beneath the seed.

Aphanomyces root rot may contribute to poor alfalfa establishment and reduced growth in wet soils. Seedlings may die (damping off) if infection occurs at an early stage of development. Older seedlings are yellowed and stunted. When *Aphanomyces* and *Phytophthora* occur together, they form a destructive disease complex. Alfalfa varieties with moderate to good levels of resistance are available for control of Aphanomyces root rot.

Anthracnose is a major cause of thinning of older alfalfa stands. Seedlings may be killed or crown and crown buds of older plants may be affected. Wilted and dead bleached stems are characteristic of the disease. Diamond-shaped lesions (cankers) with light-brown centres and dark margins develop on the lower stems.

Anthracnose is a warm, wet-weather disease. Resistant varieties are available. Anthracnose has long been recognized as a destructive disease of red clover in the southern areas. The use of resistant varieties of red clover controls this problem.

Sclerotinia crown rot occurs almost exclusively on late-summer seedings, especially when minimum-tillage methods are used. Affected plants wilt and the cottony-white mold changes to hard black bodies on the crowns or lower stems. This disease is usually seen during the cooler periods of the year. Crop rotation, with two to three years away from alfalfa, aids in control.

Verticillium wilt may be spread from field to field with infested seed and in manure from animals fed infested hay. This soilborne disease usually does not become a problem until the third production year. It can be recognized on scattered plants that become yellow and stunted, then gradually die, leaving a thin, unproductive stand. Control is mainly through use of disease-free seed, resistant varieties, and crop rotation.There are several foliage diseases of alfalfa.

Any of these may cause considerable loss of leaves

during periods of prolonged wet or humid weather. Little can be done about these diseases in the year that they occur.

SOYBEAN DISEASES

The major soybean diseases can be classified as root rots, stem rots, leaf blights, and seed diseases.

Root Rots

Several root and lower-stem rot diseases are frequently encountered. These diseases are caused by fungi that live in the soil and parasitize the lower portion of the soybean plant. The two most common are Phytophthora root rot and Rhizoctonia root rot. Phytophthora root rot is the most destructive. Thousands of acres of soybeans are destroyed by this disease each year.

The *Phytophthora* fungus attacks plants at any stage of growth, from germinating seedlings through mature plants. Infected plants generally wilt and die soon after infection or are stunted. Phytophthora root rot occurs most frequently in heavy soils with poor drainage or in years with high rainfall. Fungal populations are comprised of a number of races that attack varieties with different genes for resistance. Rhizoctonia root rot is generally more prevalent in years when weather conditions are dry in early spring and then turn wet; however the disease can occur under other conditions, as well. *Rhizoctonia* attacks the base of the plant at the soil line causing reddish-brown cankers on the lower stem.

Charcoal rot has occurred in years with high temperatures during pod-filling. Temperatures at 85 to 95 degrees F for extended periods favours disease development as does drought stress. Symptoms include stunting, yellowing of leaves, discoloured root tissues, and premature death. Diagnosis of charcoal rot is based on characteristic black zoning or charcoal-coloured speckling within root tissues. There are no adequate management methods for Rhizoctonia root rot and no resistant varieties are available for charcoal rot. Adequate fertility, a three-year crop rotation including small grains or corn may help in reducing losses.

Phytophthora root rot can be controlled with resistant soybean varieties. However, a number of different races occur, so previously resistant varieties may be susceptible to new races that develop within fields. Varieties with a high level of partial resistance are also available. These varieties are susceptible to infection, but do not have the excessive yield loss associated with susceptible varieties.

A listing of varieties with specific Rps resistance genes to the prevalent races of *Phytophthora* and ratings for the level of partial resistance to the disease. Performance of resistant and partial resistant varieties can be improved by treating seed with fungicides that control *Phytophthora.* Losses also can be reduced by using some form of tillage and providing adequate subsoil and surface drainage to problem fields.

Stem Rots

The symptoms of most stem rots develop two to three weeks prior to maturity and often are misdiagnosed as early maturing beans. Cool, wet weather conditions in midsummer favour development of Sclerotinia white mold, brown stem rot, and Diaporthe stem rot which are all found. Hot, dry summer conditions promote the development of charcoal rot. Sclerotinia white mold is the most common of the stem rot diseases and has been increasing in incidence. This disease has a characteristic fluffy white mold growth that develops on the nodes and stems.

The white mold continues to grow and causes a girdling lesion on the stem. The leaves wilt and turn brown, affected plants die, but remain erect in the field. These dead plants with wilted leaves can be spotted in fields during mid- to late August and can be confused with brown stem rot. Black, hard sclerotia (resistant fungal bodies) develop both inside and outside the stems and within the pods. These sclerotia are harvested with the beans and can be collected with seed by combines, returned back to the soil with the debris, or carried from field to field on harvesting equipment.

Management of this disease is first dependent upon preventing introduction into a field. Plant only well-cleaned seed and treat seed with a fungicide effective against *Sclerotinia*. Some varieties do appear to be less susceptible to this pathogen by having less disease by the end of the season. It will be difficult to reduce the level of *Sclerotinia* in fields with high disease severities, but long crop rotations with corn, wheat, and/or alfalfa and selecting varieties that are less susceptible should improve the productivity of affected fields.

Leaf Diseases

Several leaf diseases of soybean are common every year, but they seldom destroy enough leaf tissue to reduce yields. One of these diseases, Septoria brown spot, usually develops on the lowest leaves in late June and early July and results in a conspicuous yellowing followed by leaf-drop.

This disease is worse during wet weather and may reappear in late summer during wet, cooler weather and cause premature leaf yellowing and leaf-fall.Bacterial leaf blight occurs on leaves at various levels on the soybean plant, depending on the stage of growth in relation to the rain periods favouring disease establishment. The blighted areas of the leaves frequently drop out or tear away giving a ragged appearance to the foliage.

Downy mildew causes yellow spots on the upper sides of the leaves and grayish tufts of mold growth on the lower surfaces of leaves. Because very little or no yield loss results from the occurrence of these diseases, no control methods are recommended. Most varieties appear susceptible to these leaf diseases and continuous cropping to soybean may increase their severity.

Seed Diseases

Moldy seed at harvest results from one or more damaging diseases of soybean caused by *Phomopsis* and *Diaporthe* fungi. Phomopsis seed decay is most serious when wet conditions prevail late in the season as the seeds mature and before they dry down to a moisture level suitable for combining. Soybeans left in the field because of delayed harvest due to wet weather are almost sure to develop *Phomopsis* infections.

Moldy seed results from fungal invasion of pods and seeds. The most important aspect of Phomopsis seed decay is the effect it has on seed germination. Moldy seed will not germinate or seedlings die before emergence. Seeds that look healthy may often have seed coats colonized by *Phomopsis*. These fungi will grow into the germinating seed and kill the seedlings before emergence. The level of infected seeds can be determined by a standard germination test.

Producers interested in growing their own soybean seed should be particularly aware of the problems associated with Phomopsis seed decay. Producers should avoid growing continuous soybean and harvest those fields to be saved for seed first. Before planting, all seed should be evaluated using a standard germination test.

Use only those seed lots with 80 per cent or greater germination. On those lots with 80 to 90 per cent germination, adjust the seeding rate to compensate for the lower germination percentage, and plant only if adequate soil moisture is available for rapid germination. Most fungicide seed treatments available for use on soybean control *Phomopsis*. However, growers should not expect to increase germination of diseased seed lots by more than 20 per cent.

Soybean Cyst Nematode

Soybean cyst nematode (SCN) can be found at some level in most of the soybean producing counties of the state. Frequently no visible symptoms of nematode damage are observed other than reduced yield. Visible symptoms of SCN damage includes stunted, yellow plants and weed infestations where diseased soybeans lack adequate growth. SCN is a microscopic round worm that exists in soil as eggs, worms, and cysts.

Soils with moderate to high levels of nematodes (250 to 2,000 eggs/200 cc of soil) should implement control measures to reduce nematode populations and reduce yield losses. The only definite way to confirm the presence of SCN, or determine the level of infestation, is to submit soil samples for laboratory analysis.

WHEAT DISEASES

Many different types of diseases affect wheat. They can be classified as seed-borne diseases, leaf and head blight diseases, crown and root rot diseases, and virus diseases.

Seed-Borne Diseases

Most problems resulting from seed-borne diseases have been eliminated by highly effective seed-treatment fungicides. Several seed-borne diseases are of concern to wheat growers. They are seed-borne scab, seed-borne *Stagonospora* (previously known as *Septoria*), common bunt (stinking smut), and loose smut. Seed-borne scab and *Stagonospora* are discussed later in the section on leaf and head blight diseases.

Both diseases result in lightweight, shriveled kernels that may be moldy. Producers should follow the recommendations listed below for proper seed cleaning and planting to reduce seedling blight and stand losses resulting from planting diseased seed.

The two smut diseases, stinking smut and loose smut, can be particularly devastating. Stinking smut causes losses by giving the seed of diseased plants a foul, fishy odor, making the grain unfit for milling. Producers have been docked severely when attempting to sell smutty grain.

Loose smut affects plants and yield by converting the grain and parts of the head to smut spores. Therefore, infected plants have no grain left to harvest. There are no varieties resistant to stinking smut, and numerous races of the loose smut fungus exist. Note the following guidelines to control seed-borne diseases and seedling blights:

- Plant the highest-quality, disease-free seed possible. Seed-production fields should be inspected from the head-emergence growth stage through harvest for occurrence of scab, Stagonospora glume blotch, loose smut, and stinking smut. Do not use seed from severe smut infested fields.
- Clean seed thoroughly to remove all shriveled, lightweight kernels. This may require raising the test weight of the grain by several pounds per bushel.

- Have a standard germination test run on the seed. Use only seed with 80 per cent or greater germination percentage. If poor germination is due to Fusarium head scab, certain seed treatment fungicides can improve germination by 15 to 20 per cent.
- All wheat seed should be treated with a seed-treatment fungicide effective for control of smut fungi, Fusarium scab and *Stagonospora.*
- Plant in a well-prepared seed bed or with a no-till drill capable of proper seed placement, when soil moisture is adequate and soil temperatures are not too high. Plant after the Hessian fly safe date for your county to help avoid seedling diseases.

Leaf- and Head- Blight Diseases

Major diseases in this group are powdery mildew, leaf rust, Septoria tritici leaf blotch, Stagonospora (Septoria) nodorum leaf and glume blotch, and Fusarium head scab. All can cause major yield losses, but their occurrence is essentially weather-dependent. Cool, rainy weather from mid-April through the flowering period of the wheat plant in late May to early June favours the development of most of these diseases. They require either the leaf surfaces to be wet for a certain period of time or the relative humidity within the plant canopy to be near 100 per cent. Powdery mildew and

Septoria tritici leaf blotch are the first leaf diseases to occur in the spring. Both are favoured by cool, humid, or wet weather. Leaf rust and Stagonospora nodorum leaf blotch require slightly warmer weather, thus they follow in mid to late May. Stagonospora nodorum glume blotch and Fusarium head scab become evident in June soon after flowering, especially if wet weather persists through this time.

The use of resistant varieties is the major control procedure for these disease management, leaf, and head blight diseases. Few varieties are resistant to all of these diseases. Determine which diseases cause most consistent problems in your area, and choose a variety based on its level of resistance. The level of fungal carryover from one wheat crop to the next is minimal if a two-to three-year rotation away from wheat is maintained. Rotation is a primary control measure for powdery mildew and the *Septoria* and *Stagonospora* diseases. Spores of leaf rust are blown up from the southern states in late May; therefore, crop rotation has little effect on the incidence of leaf rust.

Adequate and balanced fertility, to provide optimum nutrition for hardy plants, helps lessen the adverse effects of foliage diseases. High rates of nitrogen will favour powdery mildew and Stagonospora leaf and glume blotch. Head scab is usually more severe when wheat is planted after corn because the fungus causing scab is the same one that causes Gibberella stalk rot. When possible, wheat should follow soybean or other legumes in the cropping sequence.

Fungicides are available to control most foliar diseases of wheat. However, the use of these fungicides should be based on sound economic decisions and the level of disease in the field. Fungicides have been profitable during years when foliar diseases have been severe on susceptible cultivars or in locations where diseases are a persistent problem.

Crown and Root Rot Diseases

Two of the more important crown- and root-rot diseases are take-all and Cephalosporium stripe. Both of these diseases are soilborne, meaning that the fungi that cause these diseases reside in the soil. Take-all can be recognized by examining the roots and lower stems of prematurely killed plants. The base of the stem and roots will have a scurfy, black appearance. Cephalosporium stripe disease is characterized by alternating yellow and brown stripes that extend the entire length of the leaf blades. Both of these diseases are favoured by planting wheat year after year in the same field.

Usually a rotation sequence with a break of two years or more between wheat crops will eliminate fungal carryover. The exception to this is when no-till is used to produce all crops in the rotation sequence and wheat residues do not decompose before the next wheat crop is planted. The other exception to this is when perennial grass weeds, such as quack grass, become established in the field. Proper rotations, tillage and elimination of grass weeds have been highly effective in managing both diseases. Additional control of the stripe disease can be achieved by maintaining a soil pH above 6.2 by proper liming according to a soil test. Adequate soil fertility will also reduce yield losses from root diseases.

Virus Diseases

Wheat spindle streak mosaic (yellow mosaic) and barley yellow dwarf are the two most common virus diseases. Wheat spindle streak mosaic is a soilborne disease that is usually recognized in early May as the stems of the wheat plant begin to elongate. The upper leaves of affected plants will show short, spindle-shaped, yellow streaks.

These symptoms may intensify if weather remains cool. The symptoms will tend to disappear as the weather begins to warm. Control of wheat spindle streak is achieved through the use of resistant varieties. A number of highly-resistant varieties are available.

Barley yellow dwarf virus is transmitted by aphids. Aphids arriving from the southern states transmit the virus to the newly-planted wheat crop in the fall. Severely affected plants may be stunted, have reddish or yellowish leaf tips and produce no heads.

Yield losses greater than 50 per cent have occurred when entire fields have been infected in the fall. Because no varieties have an acceptable level of resistance and early-fall infections cause the greatest yield losses, wheat planting should be delayed until after the Hessian fly safe date when aphids have ended their fall flights.

Chapter 6

Fruit Diseases

GRAPE BLACK ROT

Black rot is one of the most damaging grape diseases. All cultivated varieties of grapes are susceptible to infection by the black rot fungus. If not controlled, some or all of the grapes within a cluster will be rotted. The disease is favoured by warm, humid weather as is found during the summer. Before good control measures were devised, vineyards along the River often were hard hit. Grape growers commonly lost most of their crop, and the grape industry was literally driven out of the area.

Symptoms

Symptoms of black rot first appear as small yellowish spots on leaves. As the spots (lesions) enlarge, a dark border forms around the margins. The centres of the lesions become reddish brown. By the time the lesions reach 1/8 to 1/4 inch in diameter (approximately two weeks after infection), minute black dots appear.

These are fungal fruiting bodies (pycnidia) and contain thousands of summer spores (conidia). Pycnidia are often arranged in a ring pattern, just inside the margin of the lesions. Lesions may also appear on young shoots, cluster stems, and tendrils.

The lesions are purple to black, oval in outline, and sunken. Pycnidia also form in these lesions. Fruit symptoms often do not appear until the berries are about half grown. Small, round, light-brownish spots form on the fruit. The rotted tissue in the spot softens, and becomes sunken. The spot enlarges quickly, rotting the entire berry in a few days. The diseased fruit shrivels, becoming small, hard, black and wrinkled (mummies). Tiny black pycnidia are also formed on the fruit mummies. The mummies usually remain attached to the cluster.

Causal Organism

Grape black rot is caused by the fungus, *Guignardia bidwellii*. Black rot survives the winter in cane and tendril lesions and fruit mummies. In the spring during wet weather, the pycnidia on infected tissues absorb water and conidia are squeezed out. Conidia are splashed about randomly by rain and can infect any young tissue in less than 12 hours at temperatures between 60-90 degrees F. A film of water on the vine surface is necessary for infection.

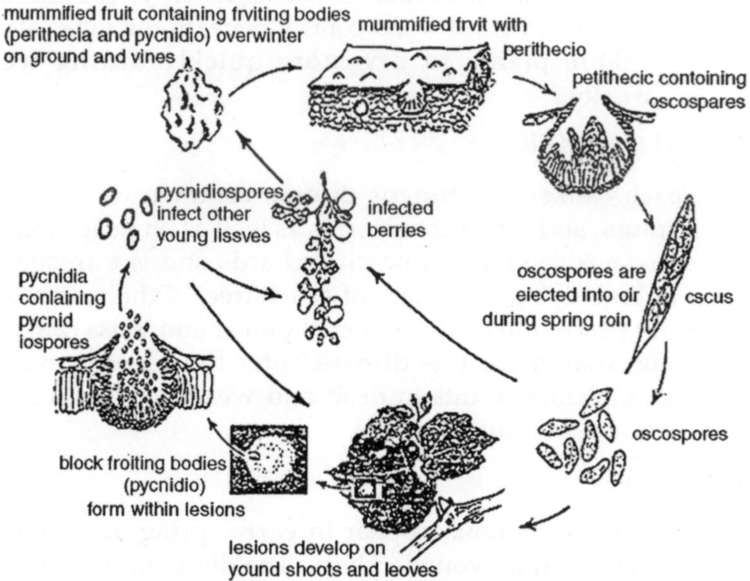

Fig. Disease Cycle of Grape Black Rot

A second type of spore, an ascospore, may also be produced in overwintered fruit mummies. Ascospores are forcibly discharged into the air and can travel considerable distances. Research has shown that ascospores are an important source of primary infections in the spring.

Control

- Sanitation is important. Destroy mummies, remove diseased tendrils from the wires, and select fruiting canes without lesions. It is very important not to leave mummies attached to the vine. Research has shown that mummies on the ground release most or all of their ascospores before the end of bloom. Mummies left up in the trellis can produce ascospores and conidia throughout the growing season, thus making control of this disease much more difficult. If only a few leaf lesions appear in the spring, remove these infected leaves.
- Plant grapes in sunny open areas that allow good air movement. Proper row orientation to prevailing winds and good weed control beneath the vines also enable plants to dry more quickly during wet weather.

PEACH CANKER

Peach canker is a fungus disease common on apricot, prune, plum, and sweet cherry trees as well as on peach trees. The disease is common in peach orchards and is a frequent cause of limb dying and death of peach trees. Other common names for peach canker are perennial canker and Valsa canker. The fungi that cause this disease enter the plant through wounds. Infection results in dead and weakened twigs and branches, and in reduced yields.

Symptoms

The first symptoms appear in early spring as gummy drops of sap around wounded bark. The diseased inner bark begins to break down, causing the cankered surface to appear depressed. Black specks, which are fungal spore producing bodies, appear on the bark surface or under the bark tissue. During wet periods spores ooze out of these fungal bodies in tiny orange or amber coloured, curled strands. During the summer, healthy bark (callus tissue) grows over the edges of the narrow, oval shaped cankers. In the fall, the fungus resumes growth, attacking and killing the new callus tissue. Over a period of years, a series of dead callus ridges form as the canker gets larger. Eventually, the canker may completely surround a branch.

The portion of the branch beyond the canker then dies. Large amounts of gum are usually produced around cankered areas. Peach canker is often confused with other problems which cause cankering and gumming. Among these are bacterial canker, insect borer injury and mechanical injury. When insects are involved, chewed-up wood dust is usually visible under the gum. Mechanical injury can often be verified by carefully reviewing recent operations in the area.

Causal Organism

Peach canker is caused by the fungi, *Cytospora leucostoma* and *Cytospora cincta*. These fungi are weak pathogens and generally do not attack healthy, vigorous peach bark. Winter injury, insect damage, and mechanical injury are common types of wounds serving as entry points. The fungi survive the winter in cankers or in dead wood.

During spring and summer, spores produced in the cankers are spread by wind and rain to wounds on the same or nearby trees. The spores are not blown over long distances in the wind. Infection and canker development depend on temperature and the species of fungus involved. *Cytospora cincta* is favoured by lower temperatures than *Cytospora leucostoma*. Because of the manner of infection and development of this disease, no single control measure is adequate. Most known control methods act indirectly by reducing points of entry or by reducing the level of inoculum. Fungicides are generally ineffective for controlling this disease.

Control

- Prune young trees carefully to avoid weak, narrow-angled crotches. Narrow-angled crotches are frequent sites of breakage and winter injury.

- Delay pruning until early spring. This promotes quick healing. Remove cankered branches and dead wood while pruning. Do not leave protruding pruning stubs. Cut flush to the next larger branch.
- Eradicate cankers and remove badly cankered limbs, branches or trees. Burn or remove all cankered limbs soon after pruning. These limbs or branches serve as a reservoir for the disease causing fungi. Sanitation is critical, especially during the early life of the orchard.
- Do not plant new peach trees near established trees with canker.
- Avoid mechanical and insect injury.
- Promote vigorous, healthy peach trees with proper fertilization, pruning, and water.
- Do not over-fertilize late in the season. Winter injury is more common on these trees because winter hardening is delayed.
- White latex paint applied to the southwest side of trunks and lower scaffold branches may help avoid cold injury.
- Maintain a good control programme for other diseases and insect pests, especially borers.

PEACH LEAF CURL

Leaf curl is a springtime disease that occurs on peach, nectarine and related ornamental plants. The disease, though not a problem every spring, can be severe during cool, wet springs that follow mild winters. The leaf curl fungus damages peach trees by causing an early leaf drop. This weakens the trees, making them more susceptible to other diseases and to winter injury. Weakened trees also will produce fewer fruit the following season. Yield may be further reduced when blossoms and young fruit become diseased and drop.

Symptoms

Symptoms of leaf curl appear in the spring. Developing

leaves become severely distorted, and have a reddish or purple cast. Later, as spores form on the leaf surface, the leaves become powdery gray in colour. Shortly after this, the leaves turn yellow or brown and drop.

There is no secondary spread of this disease from leaves infected in the spring to new leaves produced later in the growing season. Once infected leaves drop, no further symptoms will appear during that growing season. Diseased twigs become swollen and stunted, and may have a slight golden cast. They usually produce curled leaves at their tips.

Though rarely seen, flowers and fruit may also become diseased. They drop shortly after they are infected. Diseased fruit has shiny, reddish, raised, warty spots.

Causal Organism

Peach leaf curl is caused by the fungus, *Taphrina deformans*. The fungus survives the winter as spores (conidia) on bark and buds. Infection occurs very early in the growing season. During cool, wet spring weather the conidia infect new leaves as they emerge from the buds. Host plant tissues are susceptible for only a short period. As the tissues mature they become resistant. The fungus produces another type of spore (ascospore) on the upper surface of the diseased leaves.

During wet weather, ascospores produce additional conidia by budding. These conidia are carried to other parts of the tree by rain and wind, where they will overwinter until the next spring. Environment can limit leaf curl infection. This partially explains why the disease does not occur every year. Leaf curl is worse when the weather is cool and wet. Low temperatures are thought to retard maturation of leaf tissue, thus prolonging the time infection may occur. The fungus can penetrate young peach leaves readily at temperatures between 50 and 70 degrees F, but only weakly below 45 degrees F. Rain is necessary for infection.

Control

Leaf curl is not difficult to control. Since the fungus survives the winter on the surface of twigs and buds, a single fungicide spray, thoroughly covering the entire tree, will provide control. If leaf curl does result in significant defoliation in the spring, the fruit on affected trees should be thinned to compensate for the loss of leaves. Over-cropping the tree will weaken it and make it more susceptible to winter injury.

APPLE POWDERY MILDEW

Damage from powdery mildew attack results in stunted growth. The foliage becomes distorted and twig growth is reduced. Also, the fruit surface may become russetted or discoloured, and dwarfed. Heavily mildewed trees are weakened, and are more susceptible to other pests and winter injury.

Symptoms

Powdery mildew may be found on buds, blossoms, leaves, twigs, and fruit. Symptoms first appear in the spring on the lower surface of leaves, usually at the ends of branches. Small, whitish felt-like patches of fungal growth appear and quickly cover the entire leaf. Diseased leaves become narrow, crinkled, stunted and brittle. By mid-summer, tiny, black round specks show up on the lower leaf surface, but more commonly on the twigs. These are fungal fruiting bodies, but their importance in the disease cycle is probably minimal.

The fungus spreads rapidly to twigs, which stop growing and become stunted. In some cases the twigs may be killed back. Leaves and blossoms from infected buds will be diseased when they open the next spring.

Infected blossoms shrivel and produce no fruit. Fruit symptoms are not usually seen unless the disease has built up to high levels on susceptible cultivars. Diseased fruit has a fine network type surface blemish called russetting.

Causal Organism and Disease Cycle

Powdery mildew is caused by the fungus, *Podosphaera leucotricha*. Powdery mildew overwinters as fungal strands (mycelium) in vegetative or fruit buds which were infected the previous season. Infected terminals may have a silvery gray colour, stunted growth, and a misshapen appearance and are more susceptible to winter kill than are noninfected terminals. Temperatures near -18 degrees F kill a majority of mildewed buds and the fungus within them. Even at lower temperatures, however, some powdery mildew survives.

As buds break dormancy, the powdery mildew fungus resumes growth and colonizes developing shoots causing primary infections. The powdery white appearance on infected shoots consists of many thousands of spores which are responsible for spreading the fungus and causing secondary infections later in the growing season. Secondary infections are important because they produce the overwintering infected buds.

Secondary infections usually develop on leaves and buds before they harden off and may reduce the vigour of the tree. Fruitlets may become infected shortly after bloom, resulting in a web-like russetting on the mature fruit.

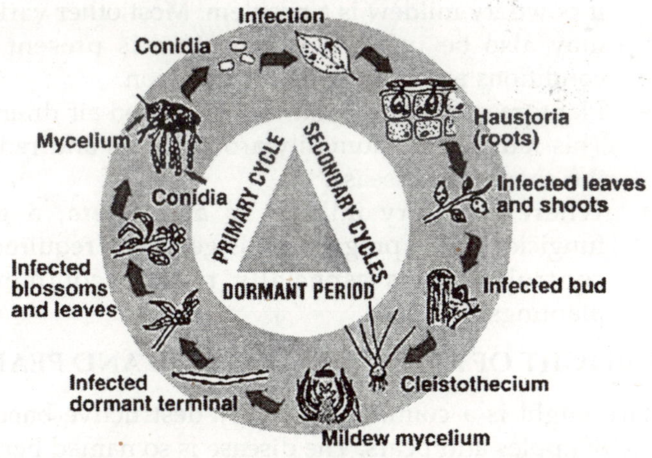

Fig. Disease Cycle of Apple Powdery Mildew

Powdery mildew infections occur when the relative humidity is greater than 90 per cent and the temperature is between 50-77 degrees F. The optimum temperature range for the fungus is 66-72 degrees F. Although high relative humidity is required for infection, the spores will not germinate if immersed in water. Leaf wetting is, therefore, not conducive to powdery mildew development. This is quite different from most other foliar and fruit diseases caused by fungi that require free water for their spores to germinate and infect.

Under optimum conditions, powdery mildew can be obvious to the naked eye 48 hr after infection. About 5 days after infection, a new crop of spores is produced. Non-germinated powdery mildew spores can tolerate hot dry conditions and may persist until favourable conditions for germination occur. It is important to remember that powdery mildew can be a problem in drier growing seasons when other diseases are not a problem.

Control

- Apple varieties vary greatly in their susceptibility to powdery mildew. Jonathan, Granny Smith, Mutsu (Crispin), Rome, Cortland, Baldwin, Monroe and Idared are very susceptible and should be avoided if powdery mildew is a problem. Most other varieties may also be infected if inoculum is present and conditions are favourable for infection.
- Plant trees in sunny locations with good air drainage. This reduces the humidity around trees and reduces the chances of disease.
- Where powdery mildew is a problem, a good fungicide spray programme is generally required for control. This is especially true in commercial plantings.

FIRE BLIGHT OF APPLES, CRABAPPLES AND PEARS

Fire blight is a common and very destructive bacterial disease of apples and pears. The disease is so named because infected leaves on very susceptible trees will suddenly turn brown, appearing as though they had been scorched by fire. As a result of this disease, blight susceptible pear cultivars are no longer grown in many parts in the Midwest.

Damage and losses from fire blight on apple result from: death or severe damage to trees in the nursery; death of young trees in the orchard; delay of bearing in young trees due to frequent blighting of shoots and limbs; loss of limbs or entire trees in older plantings as the result of girdling by fire blight cankers; and direct loss of fruit due to blighting of blossoms and young fruit. Fire blight may cause severe damage to many other members of the Rosaceae family. Quince, crabapple, mountain ash, spirea, hawthorn, pyracantha, and cotoneaster are all susceptible. Cultivars within some of these species are resistant.

Symptoms

Blossom and twig blight symptoms appear in the spring. Diseased blossoms become water-soaked and turn brown. The bacteria may then grow down into the blossom bearing twigs (spurs). Leaves on the spur become blighted, turning brown on apple and black on pear. Droplets of milky tan-coloured bacterial ooze may be visible on the surface of diseased tissue. These droplets contain millions of bacteria which can cause new infections.

Twig blight starts at the growing tips of shoots and moves down into older portions of the twig. Blighted twigs first appear water-soaked, then turn dark brown or black. Blighted leaves remain attached to the dead branches through the summer. The end of the branch may bend over, resembling a shepherd's crook or an upside down "J". As the fire blight bacteria move through blighted twigs into main branches, the bark sometimes cracks along the margin of the infected area on the main branch causing a distinct canker.

Both apple and pear fruit may be blighted. Rotted areas turn brown to black and become covered with droplets of ooze. The fruit remains firm but later dries out and shrivels into mummies.

Causal Organism

Fire blight is caused by the bacterium, *Erwinia amylovora*. The fire blight bacteria overwinter in living tissue at the margins of cankers on the trunk and main branches. The bacteria become active in the spring when temperatures get above 65 degrees F. Their growth is favoured by rain, heavy dews, and high humidity. By the time trees are blossoming, droplets of ooze containing the bacteria are present on the surface of cankers. Relatively few overwintering cankers become active and produce bacteria in the spring, but a single active canker may produce millions of bacteria, enough to infect an entire orchard.

The bacteria in droplets of ooze are spread by splashing rain or insects (mostly bees, flies, and ants) to open blossoms. The bacteria multiply rapidly in the blossom nectar, and invade the blossom tissue through natural openings called nectaries. The optimum temperature range for blossom blight infection is 65 to 86 degrees F. The bacteria are spread from blossom to blossom by rain or pollinating insects.

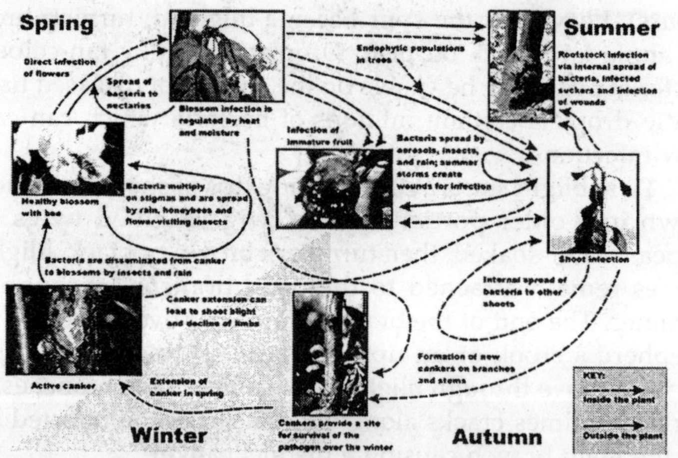

Fig. Disease Cycle of Fire Blight.

Actively growing shoot tips are infected by bacteria that have been spread by rain or insects from both cankers and infected blossoms. Invasion can occur directly through natural openings, such as lenticels and stomata, under conditions of prolonged rain and high humidity.

However, shoot infection more commonly occurs through wounds created by sucking insects, such as aphids, leafhoppers, and tarnished plant bugs; by wind whipping; or by hail. Fire blight bacteria multiply rapidly within an infected shoot. Droplets of ooze can form on the shoots within

3 days. Shoots remain highly susceptible to infection until vegetative growth ceases and the terminal bud is formed.

Control

Fireblight is one of the most difficult diseases of apple to control, and there is no one procedure that will give complete control. Though control is not an easy task, the use of several practices in an integrated manner should result in minimal damage from fire blight.

- Plant apple, crabapple, and pear varieties that are less susceptible to fire blight. Table gives a listing of the relative susceptibility of some of the more common apple and pear varieties. Fireblight is not as severe a disease problem on most crabapple varieties. A few crabapple varieties which can develop severe fireblight include: Silver Moon, Snowdrift, Red Jade, and Van Esseltine.
- Prune out fire blight cankers and blighted twigs. To decrease the inoculum level for the following season, prune out blighted twigs and cankers during the dormant season. During the dormant season (winter) there is much less chance of spreading bacteria. Branches that are more than half-girdled by cankers should be removed.

  Cut off blighted twigs by making cuts at least 4 inches below the visible dead wood. Cankers can be cut out of trunks or large branches by removing dead tissue down to wood that appears healthy. If blighted twigs are pruned out during summer, cuts should be made 12 to 15 inches below diseased wood and pruning tools should be disinfested by dipping in a 2:10 solution of household bleach in water after each cut. We recommend that commercial growers do a thorough job of pruning out blighted wood in the dormant season and not in summer.
- Follow proper pruning and fertilization practices. Excessive nitrogen fertilizer and heavy pruning will promote vigorous growth of succulent tissue which

is more susceptible to fire blight. Adjust management practices on susceptible varieties to promote moderate growth.

Make fertilizer applications in early spring or late fall after growth has ceased.

- Sucking insects create wounds through which fire blight bacteria can enter.

  These pests should be controlled throughout the growing season. To protect bees, do not apply insecticides during bloom.

SCAB OF APPLE AND CRABAPPLE

Apple scab is one of the most serious diseases of apple and ornamental crabapple. Disease development is favoured by wet, cool weather that generally occurs in spring and early summer.

Both leaves and fruit can be affected. Infected leaves may drop resulting in unsightly trees, with poor fruit production. This early defoliation may weaken trees and make them more susceptible to winter injury or other pests. Infected fruits are blemished and often severely deformed. Infected fruits may also drop early.

Symptoms

Symptoms first appear in the spring as spots (lesions) on the lower leaf surface, the side first exposed to fungal spores as buds open. At first, the lesions are usually small, velvety, olive green in colour, and have unclear margins. On some crabapples, infections may be reddish in colour.

As they age, the infections become darker and more distinct in outline. Lesions may appear more numerous closer to the mid-vein of the leaf. If heavily infected, the leaf becomes distorted and drops early in the summer. Trees of highly susceptible varieties may be severely defoliated by mid to late summer. Fruit symptoms are similar to those found on leaves. The margins of the spots, however, are more distinct on the fruit. The lesions darken with age and become black and "scabby." Scabs are unsightly, but are only skin deep. Badly scabbed fruit becomes deformed and may fall before reaching good size.

Causal Organism

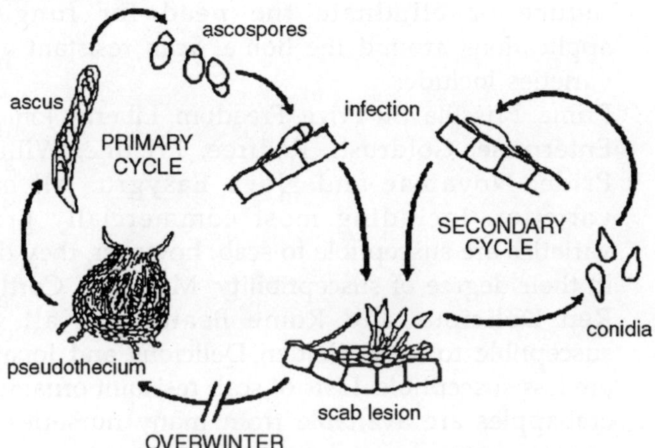

Fig. Apple Scab Disease Cycle

Apple scab is caused by the fungus, *Venturia inaequalis*. It survives the winter in the previous year's diseased leaves that have fallen under the tree. In the spring, the fungus in old diseased leaves produces millions of spores.These spores are released into the air during rain periods in April, May and June. They are then carried by the wind to young leaves, flower parts and fruits. Once in contact with susceptible tissue, the spore germinates in a film of water and the fungus penetrates into the plant. Depending upon weather conditions, symptoms (lesions) will show up in 9 to 17 days.

The fungus produces a different kind of spore in these newly developed lesions. These spores are carried and spread by splashing rain to other leaves and fruits where new infections occur. The disease may continue to develop and spread throughout the summer. Because a film of water on leaves and fruit is required for infection to occur, apple scab is most severe during years with frequent spring rains.

Control

- The use of resistant or scab immune varieties is the

ideal method for controlling scab. Currently there are several apple varieties that are totally resistant to scab. Backyard growers are strongly encouraged to consider using these resistant varieties in order to reduce or eliminate the need for fungicide applications around the home. Scab resistant apple varieties include:

Prima, Priscilla, Sir Prize, Freedom, Liberty, Jonafree, Enterprise, Goldrush, Redfree, Pristine, Williams Pride, Novamac and Nova Easygro. All other varieties, including most commercially grown varieties are susceptible to scab; however, they differ in their degree of susceptibility. McIntosh, Cortland, Red Delicious and Rome Beauty are all very susceptible to scab. Golden Delicious and Jonathan are less susceptible. Lists of scab resistant ornamental crabapples are available from many nurseries and garden centres.

- Rake and destroy fallen leaves. This will reduce the number of spores that can start the disease cycle the next year.
- Where resistance to scab is not present, fungicide application is the primary method of control.

ANTHRACNOSE OF RASPBERRY AND BLACKBERRY

Anthracnose is a disease common to raspberries, blackberries, and other brambles or cane fruits. It causes severe damage to black and purple raspberries and susceptible varieties of red raspberries throughout the United States. The disease reduces the size and quality of fruit on infected canes. In addition, it may kill canes or weaken them so that they do not survive the winter. Other common names for this disease are "cane spot" and "gray bark."

Symptoms

Anthracnose first appears in the spring on the young shoots as small, purplish, slightly raised or sunken spots. Later, they enlarge and become ash gray in the centre with slightly raised purple margins. The spots are often so close together on black and purple raspberries that they form large irregular areas (cankers).

The cankers may encircle the cane, sometimes causing the death of the cane beyond the canker. The bark in badly cankered areas often splits. Late season infections result in superficial gray, oval spots.

The spots have definite margins, but are not sunken. They may become so numerous that the spots blend together, covering large portions of the cane. This is the characteristic "gray bark" symptom which is common on red raspberry. Dark coloured specks (fungal fruiting bodies) develop in circles on the gray bark. Anthracnose sometimes attacks the leaves and can cause some leaf drop. Small spots, about 1/16 inch in diameter, with light gray centres and purple margins appear on the leaves. Lesion centres later fall out, leaving a shot hole effect.

Causal Organism

Anthracnose is caused by the fungus *Elsinoe veneta.* The fungus survives the winter in lesions on diseased canes. The following spring and summer, during wet and rainy periods, spores are released. Spores are carried by splashing rain to healthy first-year primocanes. These spores may then germinate and infect young tissues on developing primocanes. Disease development is favoured by extended periods of wet weather.

- All steps possible should be taken to improve air circulation within a planting, to allow faster drying of foliage and canes. Reducing the number and duration of wet periods should reduce the potential for infection. Excessive applications of fertilizer (especially nitrogen) should be avoided, since it promotes excessive growth of very susceptible succulent plant tissue.

  Plants should be maintained in narrow rows and thinned to improve air circulation and allow better light penetration. Weeds are very effective in reducing air movement; therefore, good weed control within and between rows is important for improving air circulation within the planting.

- After harvest, remove and destroy all old fruited canes (floricanes) and any new primocanes that are infected. It is best to remove old canes shortly after harvest, and it is critical to have them removed before new growth starts in the spring. Remember, the fungus overwinters on old-infected canes.
- Remove all wild brambles growing in the area because they can serve as a reservoir for the disease.
- Where the disease is established in the planting, fungicide applications are generally required to achieve adequate control.

SPUR BLIGHT OF RED RASPBERRIES

Spur blight is caused by the fungus *Didymella applanata*. Spur blight occurs only on red and purple raspberries. Spur blight has been considered to be a serious disease of red raspberry; however, recent studies in Scotland suggest that spur blight actually does little damage to the cane. The extent of damage caused by spur blight in the United States is not clearly understood.

The spur blight fungus has been reported to reduce yields in several ways. It can blight the fruit bearing spurs that are produced on the side branches, cause premature leaf drop, and kill buds on the canes that later develop into fruit bearing side branches. In addition, berries produced on diseased canes may be dry, small and seedy.

Symptoms

The symptoms first appear on young first-year primocanes in late spring or early summer. Purple to brown areas (lesions) appear just below the leaf or bud, usually on the lower portion of the stem.

These lesions expand, sometimes covering all the area between two leaves. In late summer or early fall, bark in the affected area splits lengthwise and small black specks, which are fungal fruiting bodies (pycnidia) appear in the lesions. They are followed shortly by many slightly larger, black, erupting spots; another form of fungal fruiting body (perithecia).

Leaflets sometimes become infected and show brown, wedge-shaped diseased areas, with the widest portion of the wedge toward the tip of the leaf. Infected leaflets may fall off, leaving only petioles without leaf blades attached to the cane. When diseased canes become fruiting floricanes during the next season, the side branches growing from diseased buds are often weak and withered.

Causal Organism

Spur blight is caused by the fungus, *Didymella applanata*. It survives the winter in lesions on diseased canes. The following spring and summer, during wet and rainy periods, spores are released and carried by splashing rain and wind to nearby primocanes. There they germinate in the presence of water and produce new infections, where the fungus will again over winter.

Control

All steps possible should be taken to improve air circulation within a planting, to allow faster drying of foliage and canes. Reducing the number and duration of wet periods should reduce the potential for infection.

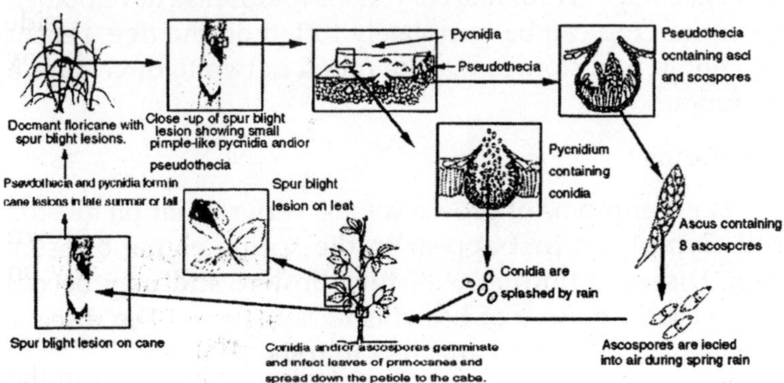

Fig. Disease Cycle of Spur Blight

Excessive applications of fertilizer (especially nitrogen) should be avoided, since it promotes excessive growth of very susceptible succulent plant tissue. Plants should be maintained in narrow rows and thinned to improve air circulation and allow better light penetration.

Weeds are very effective in reducing air movement; therefore, good weed control within and between rows is important for improving air circulation within the planting.Wild brambles, especially wild red raspberries, growing in the area should be removed. They can serve as a reservoir for the disease.

After harvest, remove and destroy all old fruited floricanes and any first-year primocanes that are infected. It is important that old canes be removed before new canes emerge in the spring. If spur blight becomes an important problem in the planting, growers may want to consider the use of fungicide. Special fungicide sprays specifically for control of spur blight are generally not warranted.

BROWN ROT OF STONE FRUITS

Brown rot is a common and destructive disease of peach and other stone fruits (plum, nectarine, apricot, and cherry). The brown rot fungus may attack blossoms, fruit, spurs (flower and fruit bearing twigs), and small branches. The disease is most important on fruits just before ripening, during and after harvest. Under favourable conditions for disease development, the entire crop can be completely rotted on the tree. Peaches not kept in cool storage may be rotted in two to three days by the fungus.

Symptoms

The symptoms of brown rot are very similar on all stone fruit. Symptoms first appear in the spring as the blossoms open. Diseased flowers wilt, turn brown, and may become covered with masses of brownish-gray spores. The diseased flowers usually remain attached into the summer.

Young fruits are normally resistant, but may become infected through wounds. As fruits mature they become more susceptible to attack, even in the absence of wounds. Fruit infections appear as soft brown spots which rapidly expand and produce a tan powdery mass of conidia. The entire fruit rots rapidly, then dries and shrinks into a wrinkled "mummy." Rotted fruit and mummies may remain on the tree or fall to the ground. Fruit infection may spread rapidly, especially if environmental conditions are favourable and fruits are touching one another.

The fungus may move from diseased blossoms or fruit into the spurs. The fungus may then invade and cause diseased areas (cankers) on the twigs below. Succulent shoots are sometimes infected by direct penetration near their tip. A canker may form encircling the twig, causing death of the twig beyond the canker (twig blight).

Causal Organism

Brown rot is caused by the fungus, *Monilinia fructicola.* The brown rot fungus survives the winter in mummified fruits (either on the ground or still on the tree) and in twig and branch cankers produced the preceding year. Both sources may produce spores that can infect blossoms and young shoots. At about blossom time, a mummified fruit that has fallen on the ground produces up to 20 or more small, tan, cup-like structures on slender stalks that are called apothecia.

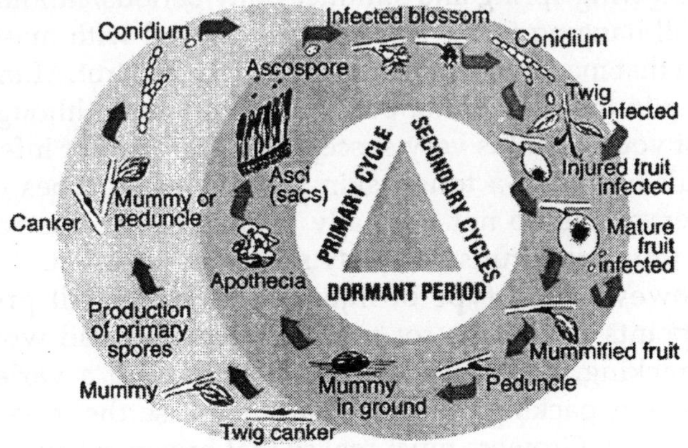

Fig. Brown Rot Disease Cycle

As an apothecium matures, it becomes thicker and the cup opens to a bowl-like disc 1/8 to 1/2 inch in diameter across the top. The inner surface of each bowl is lined with thousands of spore-containing sacs (asci). At this stage, the slightest disturbance of air movement will cause an apothecium to forcibly discharge millions of spores.

These spores (ascospores) are carried by wind to the open or unopened blossoms and young shoots. If a film of water (either from dew or rain) is present for 5 hours or longer, the spores can germinate and penetrate the plant. Infected blossoms soon wilt and tan-gray tufts, composed of masses of another type of spore (conidia), develop on the outside of the flower shuck. If the infected blossom does not drop off, the fungus soon grows through the pedicel to the twig and forms a canker.

Masses of conidia are soon produced on the newly cankered twig surface during moist periods throughout May and June. These summer spores are easily detached, and, like the ascospores, are mainly wind-borne. They are also splashed by rain or carried by insects to the growing fruit. Brown rot conidia can germinate and infect at temperatures of 32 to 90 degrees F. Wet weather and temperatures ranging from 60 to 70 degrees F are most favourable for disease development.

Following spring and summer rainy periods, mummified fruit still hanging in the tree become covered with masses of conidia that may result in blossom blight or fruit rot. Mummies hanging in the tree do not produce ascospores. Although the flesh of young fruit is very susceptible to brown rot infection, the fruit has such a tough skin that the germ tubes of the summer spores do not normally penetrate. For this reason, young uninjured fruits are fairly safe from infection.

However, any type of injury to the fruit will provide entry points for brown rot spores. Insect and hail wounds, fruit cracking, limb rubs, twig punctures, and a variety of picking and packing injuries greatly increase the losses due to brown rot. Growers must realise that brown rot spores are practically everywhere during the fruit-ripening period.

Infection is almost certain to occur if the weather is moist and if the fruit skin is broken in some way.

Control

- Sanitation is very important in controlling brown rot. All dropped and rotted fruit should be picked up and destroyed promptly. At the same time, remove all mummies from the trees. Prune out all cankers during the dormant season. Overripe or rotting fruit in the packing shed should be removed and destroyed at once.
- Control of insects that feed on fruit is essential. Remember that anything that causes wounding of the fruit will increase the incidence of brown rot. Special care should be taken during harvesting and packing to prevent puncturing or bruising of ripe fruit.
- Remove wild or neglected stone fruit trees that serve as reservoirs for the disease.
- Fruit should be cooled and refrigerated (as close to 32 degrees F as possible) immediately after harvest.
- The use of fungicide is an important part of the disease management programme for brown rot.

Orange Rust of Brambles

Orange rust is the most important of several rust diseases that attack brambles. All varieties of black and purple raspberries, and most varieties of erect blackberries and trailing blackberries are very susceptible. Orange rust does not infect red raspberries. Unlike all other fungi that infect brambles, the orange rust fungus grows "systemically" throughout the roots, crown and shoots of an infected plant, and is perennial inside the below ground plant parts. Once a plant is infected by orange rust, it is infected for life. Orange rust does not normally kill plants, but causes them to be so stunted and weakened that they produce little or no fruit.

Symptoms

Orange rust-infected plants can be easily identified shortly after new growth appears in the spring. Newly formed shoots are weak and spindly. The new leaves on such canes are stunted or misshapen and pale green to yellowish. This is important to remember when one considers control, because infected plants can be easily identified and removed at this time. Within a few weeks, the lower surface of infected leaves are covered with blister-like pustules that are waxy at first but soon turn powdery and bright orange.

This bright orange, rusty appearance is what gives the disease its name. Rusted leaves wither and drop in late spring or early summer. Later in the season, the tips or infected young canes appear to have outgrown the fungus and may appear normal. At this point, infected plants are often difficult to identify. In reality, the plants are systemically infected, and in the following years, infected canes will be bushy and spindly, and will bear little or no fruit.

Causal Organism and Disease Development

Orange rust is caused by two fungi that are almost identical, except for a few differences in their life cycles. Arthuriomyces peckianus occurs primarily in the northeastern quarter of the United States and is the causal agent for the disease. Gymnoconia nitens is a microcyclic (lacks certain spores)

Control

- Whenever possible, start with disease-free, certified nursery stock.
- When diseased plants first appear in early spring, dig them out (including roots) and destroy them before pustules form, break open, and discharge the orange masses of spores. If plants are not removed, these spores will spread the disease to healthy plants.
- Remove all wild brambles from within and around the planting site. Wild brambles serve as a reservoir for the disease.
- Maintain good air circulation in the planting by pruning out and destroying old fruited canes

immediately after harvest, thinning out healthy canes within the row, and keeping the planting free of weeds.

- Fungicide sprays are generally not considered an effective control method for orange rust.

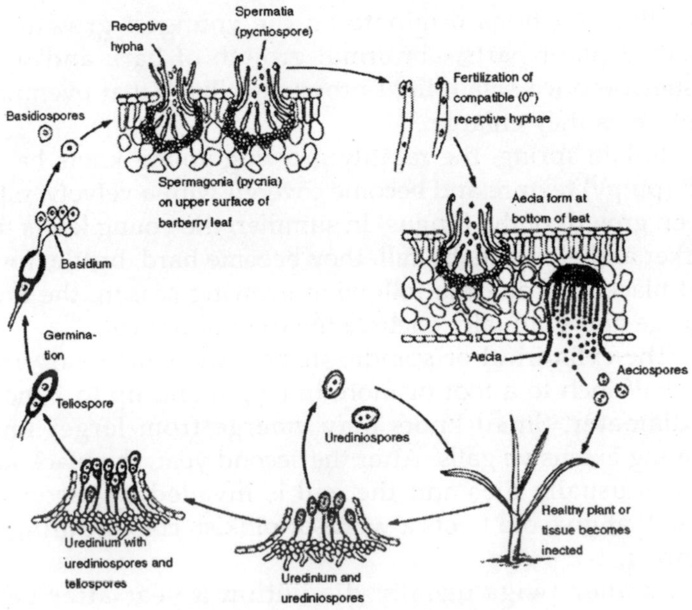

Fig. Disease cycle of turf rust

BLACK KNOT OF PLUMS AND CHERRIES

Black knot of plums and cherries is a widespread and serious disease throughout the United States. Black knot is a common disease on wild plums and cherries and in home orchards where pruning and spraying are not regularly practiced.

The disease becomes progressively worse during each growing season and unless effective control measures are taken, it can stunt or kill the tree. The black knot fungus can infect American, European, and Japanese varieties of cultivated plums and prunes. Sweet, tart, and Mahaleb cherries are also affected by the fungus, but are generally less susceptible than plum or prune. Occasionally, it may also infect apricots, peaches and other *Prunus* species.

Symptoms

The black knot fungus mainly affects twigs, branches, and fruit spurs. Occasionally, trunks may also become diseased. Usually, infections originate on the youngest growth. On infected plant parts, abnormal growth of bark and wood tissues produce small, light-brown swellings that eventually rupture as they enlarge.

In late spring, the rapidly growing young knots have a soft (pulpy) texture and become covered with a velvety, olive-green growth of the fungus. In summer, the young knots turn darker and elongate. By fall, they become hard, brittle, rough and black. During the following growing season, the knots enlarge and gradually encircle the twig or branch.

The cylindrical or spindle-shaped knots may vary from one-half inch to a foot or more in length and up to 2 inches in diameter. Small knots may emerge from larger knots forming extensive galls. After the second year, the black knot fungus usually dies and the gall is invaded by secondary fungi that give old knots a white or pinkish colour during the summer.

Smaller twigs usually die within a year after being infected. Larger branches may live for several years before being girdled and killed by the fungus. The entire tree may gradually weaken and die if the severity of the disease increases and effective control measures are not taken.

Causal Organism and Disease Development

Black knot is caused by the fungus, *Dibotryon morbosum*. The fungus overwinters in knots on twigs and branches or in the infected wood immediately surrounding them. In the spring, the fungus produces spores (ascospores) in sacs contained within tiny fruiting bodies on the surface of the knots. These ascospores are ejected into the air during rainy periods and are blown for moderate distances by wind currents.

Only succulent green twigs of the current season's growth are susceptible to infection. Ascospores that land on them may germinate and cause infection if the twigs remain wet for a sufficient length of time.

Normal growth is disrupted in the infected regions, and a knot is formed as the fungus causes the plant to produce tumorlike growths. Knots may become visible by the late summer of the year of infection but often are not noticed until the following spring, when they begin to enlarge rapidly. New ascospores capable of spreading the disease may be formed in the young knots the year following infection but often are not formed until the second spring.

The fungus continues to grow in infected wood during the spring and fall months, causing the knots to elongate several inches each year and eventually girdle affected twigs and branches.

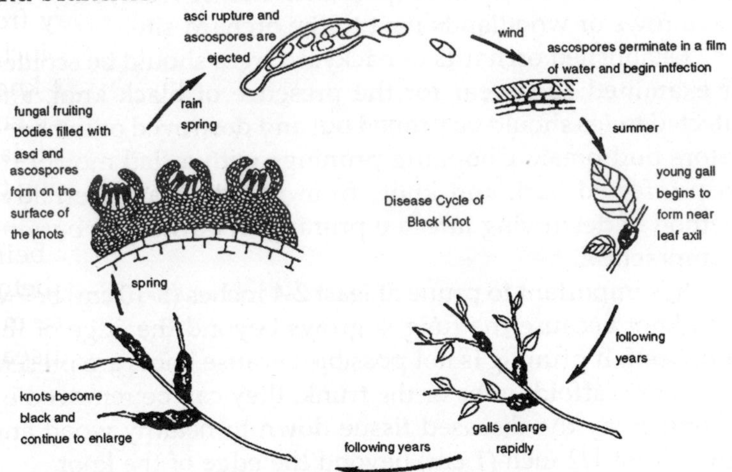

Fig. Black Knot Disease Cycle

Ascospores are potentially available from the time of bud break until terminal shoot growth stops, but the greatest number appear to be released during the period between white bud and shuck split.

Although the precise environmental conditions required for infection are uncertain, only a few hours of rain apparently are required at temperatures above 55 degrees F

(13 degrees C), whereas much longer rainy periods are required to produce infection at temperatures below this threshold.

Control

Most plum varieties grown, including Stanley and Damson, are susceptible to this disease. It has been reported that Early Italian, Brodshaw, Fallenburg, Methley and Milton are somewhat less susceptible than Stanley; Shiro, Santa Rose, and Formosa are much less susceptible; and President is apparently resistant to black knot. Japanese varieties of plums are generally less susceptible than most American varieties.

When planting new plum or prune trees, avoid planting trees next to or downwind from an old or abandoned orchard with a significant black knot problem. Similarly, remove all wild plum and cherry trees (potential disease reservoirs) from fence rows or woodlands next to the orchard site.

Established orchards or backyard trees should be scouted or examined each year for the presence of black knot, and infected twigs should be pruned out and destroyed or removed before bud break. Chopping prunings with a flail mower (to strip infected bark and knots from wood) is an alternative method of destroying infected prunings if burning or burying is impractical.

It is important to prune at least 2-4 inches (5-10 cm) below each knot because the fungus grows beyond the edge of the knot itself. If pruning is not possible because knots are present on major scaffold limbs or the trunk, they can be removed by cutting away the diseased tissue down to healthy wood and out at least 1/2 inch (1 cm) beyond the edge of the knot.

Fungicides can offer significant protection against black knot, but are unlikely to be effective if pruning and sanitation are ignored. The timing of fungicide sprays should be adjusted to account for inoculum levels and weather conditions. Where inoculum is high because of an established black knot problem or a neighbouring abandoned orchard, protection may be needed from bud break until early summer.

Where inoculum has been maintained at low to

moderate levels, sprays are most likely to be useful from white bud through shuck split (the period of maximum availability of ascospores). Fungicides are most necessary and will provide the greatest benefit if applied before rainy periods, particularly when temperatures are greater than 55 degrees F (13 degrees C). In evaluating control programmes, remember that knots often do not become apparent until the year following infection.

VERTICILLIUM WILT OF STRAWBERRY

Verticillium wilt of strawberry can be a major factor limiting production. When a plant is severely infected by the Verticillium wilt fungus, the probability of it surviving to produce a crop is greatly reduced. The Verticillium fungus can infect about 300 different host plants, including many fruits, vegetables, trees, shrubs and flowers, as well as numerous weeds and some field crops. The fungus can survive in soil, and, once it becomes established in a field or garden, it may remain alive for 25 years or longer.

Cool, overcast weather interspersed with warm, bright days is most favourable for development of Verticillium wilt. Infection and disease development may occur when soil temperature is from 70 to 75 degrees F (21 to 24 degrees C).

Many soils contain the Verticillium wilt fungus. The fungus can be introduced into uninfested soil on seed, tools and farm machinery, and in the soil and roots of transplants.

Symptoms

The first symptoms of Verticillium wilt in new strawberry plantings often appear about the time runners begin to form. In older plantings, symptoms usually appear just before picking time. Symptoms on above-ground plant parts may differ with the susceptibility of the cultivar affected. In addition, above-ground symptoms are difficult to differentiate from those caused by other root infecting fungi. Isolation from diseased tissue and culturing the fungus in the laboratory are necessary for positive disease identification.

On infected strawberry plants, the outer and older leaves

droop, wilt, turn dry and become reddish-yellow or dark brown at the margins and between veins. Few new leaves develop, and those that do tend to be stunted and may wilt and curl up along the midvein. Severely infected plants may appear stunted and flattened, with small yellowish leaves. Brownish to blueish-black streaks or blotches may appear on the runners or petioles. New roots that grow from the crown are often dwarfed with blackened tips. Brownish streaks may occur within the decaying crown and roots.

If the disease is serious, large numbers of plants may wilt and die rapidly. When the disease is not so serious, an occasional plant or several plants scattered over the entire planting may wilt and die.

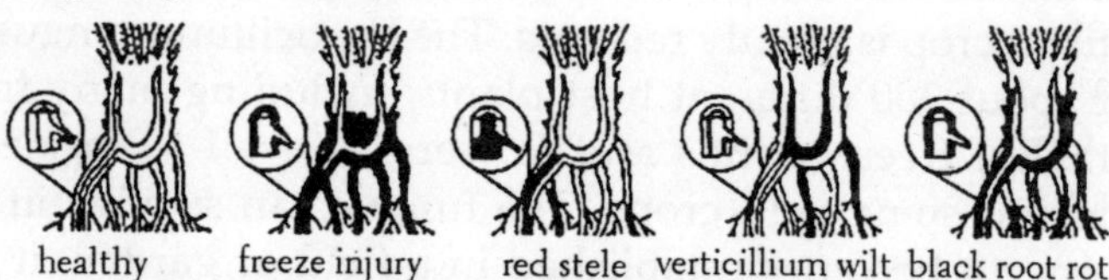

Fig. Some Common Strawberry Root Problems and Typical Symptoms

Casual Organism

Verticillium wilt is caused by the soilborne fungus fumigant that is most effective is a mixture of chloropicrin (tear gas) 33 per cent and methyl bromide 66 per cent.

DOWNY MILDEW OF GRAPE

Downy mildew is a major disease of grapes throughout the eastern United States. The fungus causes direct yield losses by rotting inflorescences, clusters and shoots. Indirect losses can result from premature defoliation of vines due to foliar infections. This premature defoliation is a serious problem because it predisposes the vine to winter injury. It may take a vineyard several years to fully recover after severe winter injury.

Symptoms

On leaves, young infections are very small, greenish-yellow, translucent spots that are difficult to see. With time the lesions enlarge, appearing on the upper leaf surface as irregular pale-yellow to greenish-yellow spots up to 1/4 inch or more in diameter. On the underside of the leaf, the fungus mycelium (the "downy mildew") can be seen within the border of the lesion as a delicate, dense, white to grayish, cotton-like growth. Infected tissue gradually becomes dark brown, irregular, and brittle.

Severely infected leaves eventually turn brown, wither, curl, and drop. The disease attacks older leaves in late summer and autumn, producing a mosaic of small, angular, yellow to red-brown spots on the upper surface. Lesions commonly form along veins, and the fungus sporulates in these areas on the lower leaf surface during periods of wet weather and high humidity.

On fruit, most infection occurs during 2 distinct periods in the growing season. The first is when berries are about the size of small peas. When infected at this stage, young berries turn light brown and soft, shatter easily, and under humid conditions are often covered with the downy-like growth of the fungus. Generally, little infection occurs during hot summer months. As nights become cooler in late summer or early fall, the second infection period may develop.

Berries infected at this time generally do not turn soft or become covered with the downy growth. Instead, they turn dull green, then dark brown to brownish-purple. They may wrinkle and shatter easily and, in severe cases, the entire fruit cluster may rot. These infected fruit will never mature normally. On shoots and tendrils, early symptoms appear as water-soaked, shiny depressions on which the dense downy mildew growth appears. Young shoots usually are stunted and become thickened and distorted. Severely infected shoots and tendrils usually die.

Causal Organism

Downy mildew is caused by the fungus *Plasmopara viticola.* The fungus overwinters in infected leaves on the ground and possibly in diseased shoots. The overwintering spore (oospore) germinates in the spring and produces a different type of spore (sporangium). These sporangia are spread by wind and splashing rain. When plant parts are covered with a film of moisture, the sporangia release small swimming spores, called zoospores. Zoospores, which also are spread by splashing rain, germinate by producing a germ tube that enters the leaf through stomates (tiny pores) on the lower leaf surface.

The optimum temperature for disease development is 64 to 76 degrees F (18 to 25 degrees C). The disease can tolerate a minimum temperature of 54 to 58 degrees F (12 C to 13 degrees C), and a maximum temperature of about 86 degrees F (30 degrees C). Once inside the plant, the fungus grows and spreads through tissues. Infections are usually visible as lesions in about 7-12 days.

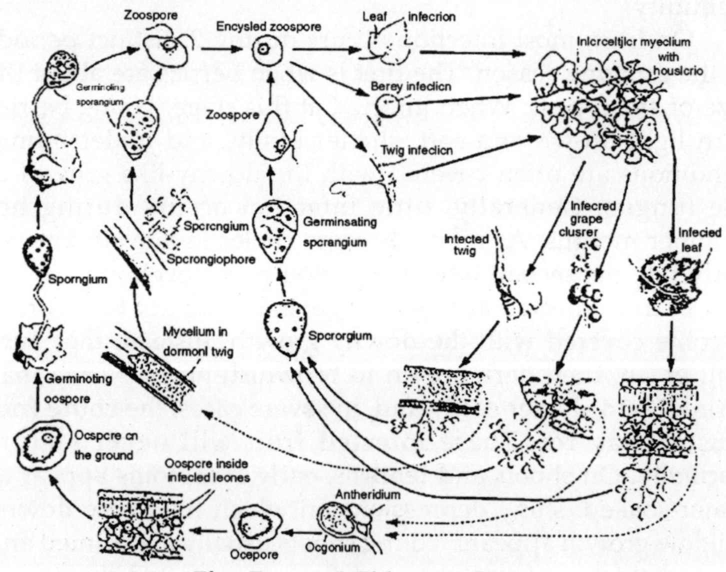

Fig. Downy Mildew on Grape

At night during periods of high humidity and temperatures above 55 degrees F (13 degrees C), the fungus grows out through the stomates of infected tissue and produces microscopic, branched, tree-like structures (sporangiophores) on the lower leaf surface. More spores (sporangia) are produced on the tips of these tree-like structures. The small sporangiophores and sporangia make up the cottony, downy mildew growth. Sporangia cause secondary infections and are spread by rain.

Control

Any practice that speeds the drying time of leaves and fruit will reduce the potential for infection. Select a planting site where vines are exposed to all-day sun, with good air circulation and soil drainage. Space vines properly in the row, and, if possible, orient the rows to maximize air movement down the row.

Sanitation is important. Remove dead leaves and berries from vines and the ground after leaf drop. It may be beneficial to cultivate the vineyard before bud break to cover old berries and other debris with soil. Cultivation also prevents overwintering spores from reaching developing vines in the spring.

To improve air circulation, control weeds and tall grasses in the vineyard and surrounding areas. When pruning, select only strong, healthy, well-coloured canes of the previous year's growth. Practices such as shoot positioning and leaf removal that help to open the canopy for improved air circulation and spray coverage are also very important.

Grape varieties vary greatly in their susceptibility to downy mildew. In general, *vinifera* (*Vitis vinifera*) varieties are much more susceptible than American types, and the French hybrids are somewhat intermediate in susceptibility. Cabernet Franc, Cabernet Souvignon, Catawba, Chancellor, Chardonnay, Delaware, Fredonia, Gewurytraminer, Ives, Merlot, Niagra, Pinot Blanc, Pinot Noir, Riesling, Rougeon and Sauvignon Blanc are reported to be highly susecptible to downy mildew.

A good fungicide spray programme is extremely important. Downy mildew can be effectively controlled by properly timed and effective fungicides.

RED STELE ROOT ROT OF STRAWBERRY

Many commercial strawberry cultivars are susceptible to

the red stele fungus. This root rot disease has become a serious problem facing strawberry production in the northern two-thirds of the United States. The disease is most destructive in heavy clay soils that are saturated with water during cool weather when the fungus is most active. The red stele fungus can survive in soil for up to 13 years or longer once it becomes established in the field or garden.

Normally, the disease is prevalent only in the lower or poorly drained areas of the planting; however, it may become fairly well distributed over the entire patch, especially during a cool, wet spring.

Symptoms

When plants start wilting and dying in the lower portions of the strawberry planting, the cause is very likely to be red stele. Infected plants are stunted, lose their shiny-green luster, and produce few runners. Younger leaves often have a metallic, blueish-green cast. Older leaves turn prematurely yellow or red. With the first hot, dry weather of early summer, diseased plants wilt rapidly and die.

Diseased plants have very few new roots, when compared with the roots of healthy plants that have thick and bushy roots with many secondary feeding roots. Infected strawberry roots usually appear gray, while the new roots of a healthy plant are yellowish-white.

The most reliable symptom of red stele is found within the roots and may be observed by gently digging up a few plants that are just beginning to wilt, taking care to preserve the root system. Plants with red stele usually have few fine lateral roots so the main fleshy roots have a "rat-tail" appearance. During intermediate stages of disease development, these fleshy roots will be white near the crown of the plant but will show a dark rot progressing upward from the tips.

When the white outer portion of the root just above this rotten zone is peeled off or sliced through, the root core (or stele) will appear to be dark red. It may be necessary to examine several rotting roots before finding a red stele, but this symptom is very distinctive and is diagnostic for the disease. Reddened steles are relatively difficult to find after harvest because most infected roots have died and begun to decay by then.

Causal Organism and Disease Cycle

Red stele is caused by the soilborne fungus *Phytophthora fragariae*. This fungus is not a natural inhabitant of most agricultural soils but probably is introduced on nursery stock or by the movement of infested soil and runoff water from fields in which the disease occurred previously.

P. fragariae is very persistent and can survive in a field for many years once it has become established, even if no strawberries are grown during that time. The organism that causes red stele of strawberry is not known to cause disease on any other crop, with the possible exception of loganberry.

P. fragariae persists in the soil as thick-walled resting spores (oospores). When the soil is moist or wet, some of the oospores germinate and form structures called sporangia, which are filled with the infectious spores of the fungus (zoospores). These microscopic zoospores are released into the soil when it becomes completely saturated with water (flooded or puddled) and use tail-like structures to swim short distances through water-filled soil pores to the tips of strawberry roots, to which they are chemically attracted.

Zoospores also may swim to the soil surface, where surface runoff water may carry them relatively long distances. Zoospore activity may occur at soil temperatures ranging from about 38 to 77 degrees F (4 to 25 degrees C), but is most significant from 44 to 59 degrees F (7 to 15 degrees C). Thus infection is most likely in the spring and fall.

Once zoospores have infected the root tip, the fungus begins to grow up into other parts of the root, causing the characteristic dark rot and red stele symptoms. New sporangia are formed along the outside of infected root issue and release additional zoospores whenever the soil is saturated, thereby continuing to spread the disease. The fungus produces oospores within infected roots as they begin to rot and die, and these oospores are released into the soil when the roots decay, thus completing the disease cycle.

Control

Since significant production and movement of infective zoospores occurs only during periods when the soil is completely saturated, the key to control is drainage. Strawberries should not be planted in low-lying or heavy soils where water accumulates or is slow to drain. On marginal soils, planting strawberries on beds raised at least 10 inches high will bring much of the root system above the zone of greatest pathogen activity and the severity of red stele root rot should be significantly reduced.

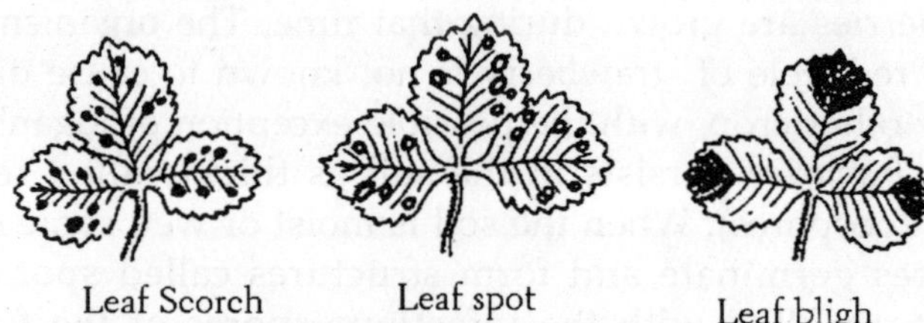

Fig. Typical Symptoms of Leaf scorch, Spot, and Blight on Strawberry Leave

Strawberry varieties highly resistant to red stele should be seriously considered for planting in a marginally drained site or a field in which red stele has been suspected of occurring in the past. Only resistant varieties should be planted in a field where red stele is known to have caused losses within the last 5 to 10 years. The following junebearing varieties are reported to be resistant to Red Stele:

Allstar; Delite; Earliglow; Guardian; Lester; Midway; Redchief; Scott; Sparkle; Sunrise and Surecrop. The everbearing varieties are also reported to be resistant. All "resistant" varieties, however, are resistant only to certain common races of the red stele fungus and can become diseased if exposed to other races of the pathogen.

It is important to minimize the chance of introducing the red stele fungus into a field where it does not already exist. Buy nursery stock only from a reputable supplier, and take care not to transfer soil on farm implements from an infested field into a clean one. New fungicides active against red stele also help in controlling this disease but are most effective when used in combination with good soil water management practices.

STRAWBERRY LEAF SPOT DISEASES

Fungal diseases of the leaf may occur as soon as the first leaves unfold in early spring and continue until dormancy in the late fall. Generally, these diseases do not cause significant economic damage. The primary damage from leaf diseases is a loss of vigour through reduced leaf area. If outbreaks of these leaf diseases become significant, the plants will become weakened resulting in increased susceptibility to root diseases and winter injury.

The three major leaf diseases that are caused by fungi have a similar disease cycle and are controlled in a similar manner. Leaf spot, leaf scorch, and leaf blight are the most common leaf diseases and they all overwinter in infected dead or living leaves. They all produce spores that spread the disease by causing new infections during moist, warm conditions.

Leaf Spot

Leaf spot is caused by the fungus, *Mycosphaerella fragariae*. Symptoms of leaf spot first appear as circular, deep purple spots on the upper leaf surface. These spots enlarge and the centres turn grayish to white on older leaves and light brown on young leaves. A definite reddish purple to rusty brown border surrounds the spots.

On fruit, superficial black spots may form under moist weather conditions. The spots form on ripe berries around groups of seeds. The spots are about 1/4 inch in diameter, and there are usually only one or two spots per fruit. However, some fruits may be more severely infected. The fungus overwinters as spores in lesions on leaves. The fungus produces more spores in spots on the upper and lower leaf surface that spread the disease during early summer. These spores are spread by splashing rain.

Middle-aged leaves are most susceptible. Lesions also develop on stems, petioles and runners.

LEAF SCORCH

Leaf scorch is caused by the fungus *Diplocarpon earliana*. Symptoms of leaf scorch consist of numerous small, irregular, purplish spots or "blotches" that develop on the upper surface of leaves. The centres of the blotches become brownish. Blotches may coalesce until they nearly cover the leaflet, which then appears purplish to reddish to brown. The fungus overwinters on infected leaves.

The fungus produces spore forming structures in the spring on both surfaces of dead leaves. These structures produce spores abundantly in midsummer. In the presence of free water these spores can germinate and infect the plant within 24 hrs. Older and middle-aged leaves are infected more easily than young ones.

LEAF BLIGHT

Leaf blight is caused by the fungus, *Phomopsis obscurans*. Symptoms of leaf blight infections begin as one to several circular reddish-purple spots on a leaflet. Spots enlarge to V-shaped lesions with a light brown inner zone and dark brown outer zone. Lesions follow major veins progressing inward. The whole leaflet may turn brown. In severe cases, stolons, fruit trusses and petioles may become infected which may girdle and kill the stem.

The fungus overwinters as mycelium or fruiting structures on the old leaves that remain attached to the plant. Spores are spread by rain splash early in the spring. Leaf blight is most destructive to older leaves in the late summer. Petioles, calyxes and fruit may also be infected earlier in the season.

Control

Leaf spot and leaf scorch are controlled most effectively by the use of resistant varieties. The following junebearing varieties are reported to be resistant to both leaf spot and leaf scorch: Allstar; Canoga; Cardinal; Delite; Earliglow; Honeoye;

Jewell; Lester; Midway and Redchief. The ever bearing varieties, Tribute and Tristar, are reported to be tolerant to leaf spot and leaf scorch. There are no varieties with reported resistance to leaf blight. These cultural practices should help reduce infection:

- Remove the older and infected leaves from runner plants before setting.
- Take care in spacing runner plants in matted-row culture.
- Plant in light, well-drained soil in a location exposed to all-day sun and good air circulation.
- Control weeds in the planting. Weeds reduce air circulation and increase drying time for leaves. (Leaves stay wet longer in weedy plantings.)
- Removing infected leaves after harvest (during renovation) is helpful in reducing inoculum and controlling all the leaf diseases.

If leaf diseases are a problem in the planting, fungicides will aid in control.

BOTRYTIS FRUIT ROT "GRAY MOLD"

Many fungi are capable of rotting mature or near-mature fruits of strawberry, raspberry, and blackberry. Under favourable environmental conditions for disease development, serious losses can occur. One of the most serious and common fruit rot diseases is gray mold. The gray mold fungus can affect petals, flower stalks (pedicels), fruit caps, and fruit. In wet, warm seasons, probably no other disease causes a greater loss of flowers and fruit. The disease is most severe during years with prolonged rainy and cloudy periods during bloom or during harvest.

Symptoms

Young blossoms are usually very susceptible to infection. One or several blossoms in a cluster may show blasting (browning and drying) that may extend down the pedicel. Fruit infections usually appear as soft, light brown, rapidly enlarging areas on the fruit. If infected fruits remain on the plant, the berry usually dries up, "mummifies," and becomes covered with a gray, dusty powder, which gives the disease its name "gray mold." Fruit infection is most severe in well-protected areas of the plant, where the humidity is high and air movement is poor. On strawberry, berries resting on soil or touching another decayed berry or a dead leaf in dense foliage are most commonly affected.

The disease may develop on young green fruits, but fruits become more susceptible as they mature. Usually, the disease is not detected until fruits are mature at harvest time. After picking, mature fruits are extremely susceptible to gray mold, especially if bruised. During picking, the handling of infected fruit will spread the fungus to healthy ones. Under favourable conditions for disease development, healthy berries may become a rotted mass within 48 hours after picking.

Causal Organism

Gray mold is caused by the fungus *Botrytis cinerea*. The fungus is capable of infecting a great number of different plants. The disease cycle is very similar for both strawberries and brambles. The fungus overwinters as minute, black, fungus bodies (sclerotia) or as mycelium in plant debris, such as dead strawberry or raspberry leaves.

Fig. Gray Mold on Strawberry

Recent research has shown that nearly all of the overwintering inoculum in strawberry plantings comes from mycelium in dead strawberry leaves within the row or planting. In early spring, the mycelium becomes active and produces large numbers of microscopic spores (conidia) on the surface of old plant (leaf) debris in the row.

Spores are spread by wind throughout the planting where they are deposited on blossoms and fruits. They germinate when a film of moisture is present and infection can occur within a few hours. Temperatures between 70 and 80 degrees F (20 to 27 degrees C) and free moisture on the foliage from rain, dew, fog, or irrigation water are ideal conditions for disease development.

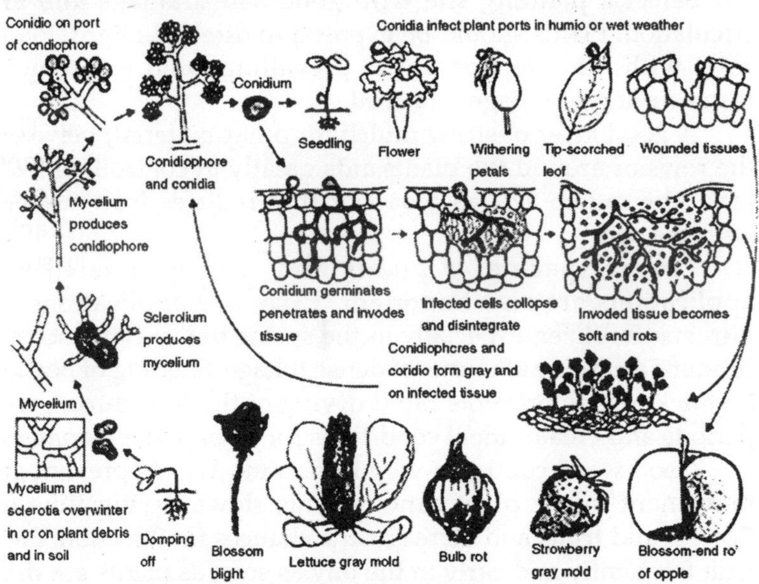

Fig. Botrytis Gray Mold on Raspberry Fruit

The disease can develop at lower temperatures if foliage remains wet for long periods. Strawberries and raspberries are susceptible to Botrytis during bloom and again as fruits ripen. Recent research indicates that most fruit infection actually occurs during bloom; however, symptoms usually do not develop until close to harvest. During bloom, the fungus colonizes healthy or senescing flower parts, often turning the blossoms brown.

These blossom infections establish the fungus within the receptacle of the young fruit as a "latent" or "quiescent" infection. The fungus generally remains latent in developing (green) fruit until the fruit starts to mature, at which time the fungus becomes active and symptoms (rot) appear. Thus, the most critical perio d for applying fungicides to control gray mold is during bloom. This is an important point to remember when considering fungicide applications for controlling this disease.

Control

Select a planting site with good soil drainage and air circulation. Plants should be exposed to direct sunlight. Plant rows with the direction of the prevailing wind to promote faster drying of foliage and fruit.

A good layer of straw mulch (or other material) between the rows or around the plants aids greatly in controlling fruit rots. The mulch acts as a barrier that reduces fruit contact with the soil.

Proper spacing of plants and timing of fertilizer applications are also important. Excessive applications of nitrogen fertilizer, especially in the spring before harvest, can produce excessive amounts of dense foliage. Shading of berries by thick foliage prevents rapid drying of the fruit during wet periods and creates ideal conditions for disease development.

Good weed control is very important. Weeds prevent air movement in the plant canopy. This slows drying time of flowers and fruits and increases the chances for infection. Pick fruit frequently and early in the day as soon as plants are dry. Cull out all diseased berries but do not leave them in the field. Handle berries with care to avoid bruising. Refrigerate fruit promptly at 32 to 50 degrees F to check gray mold.

Fungicides are an important disease management tool in commercial plantings, but are generally not effective unless they are timed properly and used in conjunction with the above mentioned cultural practices. Homeowners are encouraged to emphasize the use of cultural practices in order to avoid the use of fungicides.

POWDERY MILDEW OF GRAPE

Powdery mildew is an important disease of grapes throughout. The disease generally is considered less economically important than black rot or downy mildew. However, uncontrolled, the disease can be devastating on susceptible varieties under the proper environmental conditions. Unlike black rot and downy mildew, the powdery mildew fungus does not require free water on the plant tissue surface to infect. Powdery mildew can result in reduced vine growth, yield, fruit quality, and winter hardiness. Varieties of *Vitis vinifera* and its hybrids generally are much more susceptible than American varieties.

Symptoms

The powdery mildew fungus can infect all green tissues of the vine. Small, white or grayish-white patches of fungal growth appear on the upper or lower leaf surface. These patches usually enlarge until the entire upper leaf surface has a powdery, white to gray coating. The patches may remain limited throughout most of the season. Severely affected leaves may curl upward during hot, dry weather. Expanding leaves that are infected may become distorted and stunted.

On young shoots, infections are more likely to be limited, and they appear as dark-brown to black patches that remain as dark patches on the surface of dormant canes.

If blossom clusters are affected, the flowers may wither and drop without setting fruit. Infections on cluster stems often go unnoticed, but can be very damaging. Infected cluster stems may wither and dry up, resulting in berry drop (shelling). Affected berries may have spots on the surface similar to those on the leaves, or the entire berry may be covered with the white, powdery growth. Infected berries often are misshapen or have rusty spots on the surface.

Severely affected fruit often split open. When berries of purple or red cultivars are infected as they begin to ripen, they fail to colour properly and have a blotchy appearance at harvest. Berries are susceptible to infection until their sugar content (0Brix) reaches about 8 per cent.

Late in the season, many black specks may develop on the surface of infected areas. These are the sexual fruiting bodies (cleistothecia) of the fungus.

Causal Organism and Disease Cycle

Powdery mildew is caused by the fungus *Uncinula necator*. It was previously thought that the powdery mildew fungus overwintered inside dormant buds of the grapevine. Recent research has shown that almost all overwintering inoculum in comes from cleistothecia, which are fungal fruiting bodies that overwinter primarily in bark crevices on the grapevine.

In the spring, airborne spores (ascospores) released from the cleistothecia are the primary inoculum for powdery mildew infections. Ascospore discharge from cleistothecia is initiated if 0.1 inch of rain occurs with an average temperature of 50 degrees F. Most mature ascospores are discharged within 4-8 hours. Ascospores are carried by wind.

They germinate on any green surface on the developing vine, and enter the plant resulting in primary infections. The fungus produces another type of spore (conidia) over the infected area after 6-8 days. The conidia and fungus mycelia on which they are formed give the powdery or dusty appearance to infected plant parts.

The conidia serve as "secondary inoculum" for powdery mildew infection throughout the remainder of the growing season. It is important to note that a primary infection caused by one ascospore can result in the production of hundreds of thousands of conidia, each of which is capable of causing secondary infections that spread the disease.

Temperatures of 68-77 degrees F are optimal for infection and disease development, although infection can occur from 59-90 degrees F. Temperatures above 95 degrees F inhibit germination of conidia and above 104 degrees F they are killed. High relative humidity is conducive to production of conidia. Atmospheric moisture in the 40-100 per cent relative humidity range is sufficient for germination of conidia and infection. Free moisture, especially rainfall, is detrimental to survival of conidia.

This is in direct contrast to most other grape diseases such as black rot and downy mildew that require free water on the plant surface before the fungus spores can germinate and infect. Low, diffuse light seems to favour powdery mildew development.

Under optimal conditions, the time from infection to production of conidia is about 7 days. It is important to remember that powdery mildew can be a serious problem in drier growing seasons when it is too dry for other diseases such as black rot or downy mildew to develop.

Cleistothecia are formed on the surface of infected plant parts in late fall. Many of them are washed into bark crevices on the vine trunk where they overwinter to initiate primary infections during the next growing season.

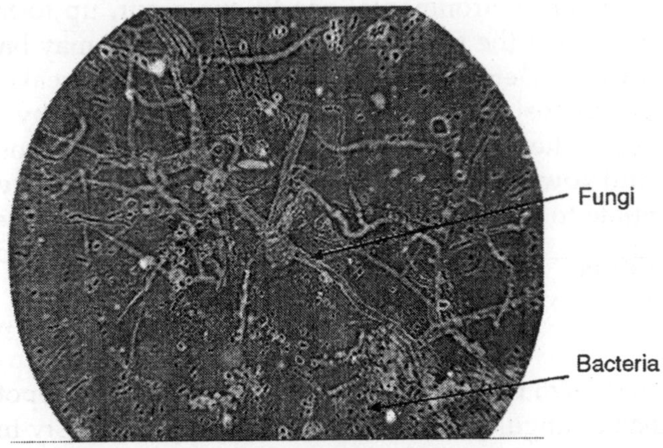

Fig. Grape Powdery

Control

Select an open planting site with direct sunlight. Plant rows in the direction of the prevailing wind in order to promote good air circulation and faster drying of foliage and fruit. Prune and train vines properly in such as way as to reduce shading and increase air circulation.

Varieties differ greatly in their susceptibility to powdery mildew. Cabernet Franc, Cabernet Sauvigon, Chancellor, Chardonnay, Chelois, Gewurztraminer, Merlot, Pinot blanc, Pinot noir, Riesling, Rosette, Rougeon, Sauvignon blanc, Seyval, Vidal 256 and Vignoles are all highly susceptible.

On susceptible varieties, control is based on the use of properly timed applications of effective fungicides. Early season (prebloom through bloom) control of primary infections caused by ascospores must be emphasized.

BACTERIAL SPOT OF STONE FRUITS

Bacterial spot affects peaches, nectarines, apricots, plums, prunes, and cherries. The disease is widespread throughout all fruit growing states east of the Rocky Mountains. Bacterial spot can affect leaves, twigs, and fruit. Severe infection results in reduced fruit quality and yield. Fruit infection is most serious on late-maturing varieties.

If proper environmental conditions occur, up to 50 per cent or more of the fruit of susceptible varieties may have to be discarded. Defoliation by bacterial spot may weaken the tree, predisposing it to winter injury and attack by other pathogens. The disease is usually more severe where soils are light and low in fertility. Vigorous trees are usually less susceptible to the disease than devitalized, neglected trees.

Symptoms

Leaves

Small (1/25 to 1/5 inch) spots form in the leaves. Spots are irregular to angular and have a deep purple to rusty-brown or black colour. In time, the centres dry and tear away leaving ragged "shot-holes." When several spots merge, the leaf may appear scorched, blighted, or ragged. Badly infected leaves may turn yellow and drop early. Early defoliation is most common on trees deficient in nitrogen or where the disease is further complicated by pesticide injury.

Nitrogen deficiency can cause leaf symptoms that are very similar to those of bacterial spot. Care must be taken to avoid an incorrect diagnosis. In both cases, the leaves of affected trees turn yellow and drop prematurely. With bacterial spot, symptoms are most likely to appear at the tip of the leaf but with nitrogen deficiency, symptoms are usually most evident along the midrib. Leaf tissues surrounding shot-holes caused by a nitrogen deficiency are more likely to have a reddish colour than when bacterial spot is involved.

Fruit

Small, round olive-brown to black spots form on the fruit. They are usually sunken and frequently surrounded by a water-soaked margin. On peaches, spots usually form on the side exposed to the sun. Spots may slowly enlarge and merge to cover large irregular areas on the fruit. On some varieties the spots may exude a yellowish gum after rainy periods. Skin cracking and pitting may occur near the spots during fruit enlargement. Fruit infected at an early stage of development are usually the most malformed.

On plums, symptoms are different than on peaches. Large, sunken, black spots form on some varieties; on others, small pit-like lesions are common.

Twigs

On peaches, two distinct types of cankers damage twigs. "Spring cankers" develop on young twigs produced the previous summer. Spring cankers first appear as water-soaked, slightly darkened blisters about the time the first leaves appear. If these cankers encircle the twig, it will die. As the season progresses, the tissues over the blister-like lesions rupture and bacteria are released. These bacteria can be spread by windblown or splashing rain and can result in new infections. In time, spring cankers heal and become inactive.

"Summer cankers" develop on green twigs of the current seasons growth. They usually occur later in the summer after leaf spots are evident. At first, they are water-soaked, dark purplish spots. In time, they enlarge, turn brown to purple-black, become slightly sunken, and are round to elliptical with water-soaked margins.

On certain plum and apricot varieties, twig cankers may continue to develop in two- and three-year old twigs. If the cankers are deep-seated, they can deform or kill the twigs.

Causal Organism

Bacterial spot is caused by the bacterium Xanthomonas pruni. The bacteria overwinter in twigs that are infected late in the season about the time leaves are shed. The following spring, when environmental conditions are favourable, bacteria ooze out onto the surface of these twigs.

The bacteria are then spread by windblown or splashing rain and can result in new infections throughout the growing season. The bacteria come in contact with healthy leaves, fruit, and current-year twigs and enter the tissues through stomata or lenticels when surface moisture is present. Once inside healthy tissues, the bacteria multiply and disease develops.

Warm temperatures (70-85 degrees F, 21-29 degrees C) with light rains, heavy dews, and windy weather are most conducive for disease development and spread. The disease makes little progress when weather is hot and dry.

Control

- When planning an orchard, avoid low-lying or shaded sites with poor air circulation and soil drainage.
  Any practice that promotes faster drying of fruit and foliage will help reduce the risk of infection. Destroy nearby wild or neglected stone fruits (Prunus species). Buy and plant only vigorous, disease-free fruit trees from a reputable nursery.
- Prune trees annually to allow for better air circulation and to maintain tree vigour. If possible, prune during dry weather in the latter half of the dormant season.
- Select peach varieties with resistance to bacteria spot. The following varieties are somewhat resistant: Belle of Georgia, Biscoe, Candor, Comanche, Garnet Beauty, Harbrite, Harken, Late Sunhaven, Loring, Madison, Norman, Pekin, Raritan Rose, Redhaven,

Redskin and Sunhaven. These varieties are very susceptible: Babygold S, Blake, Elberta, Halehaven, Jersey Queen, Jerseyland, July Elberta, J.H. Hale, Kalhaven, Rio-Oso-Gem, Suncling, Suncrest, and Sunhigh.

- Fertilize where needed to maintain vigorous but not excessive shoot growth.
- Spray applications. At present, no spray programme is completely effective for controlling bacterial spot. In the home orchard, spraying for bacterial spot is not considered practical.
- Scab of Peach, Nectarine, Plum, and Apricot

Scab occurs throughout the Midwest, wherever peaches, nectarines, plums, and apricots are grown. The disease affects fruit, leaves, and young green twigs. Scab is most common in home orchards where fungicide spray programmes are not practiced. The general use of fungicides by commercial growers has greatly reduced losses from scab.

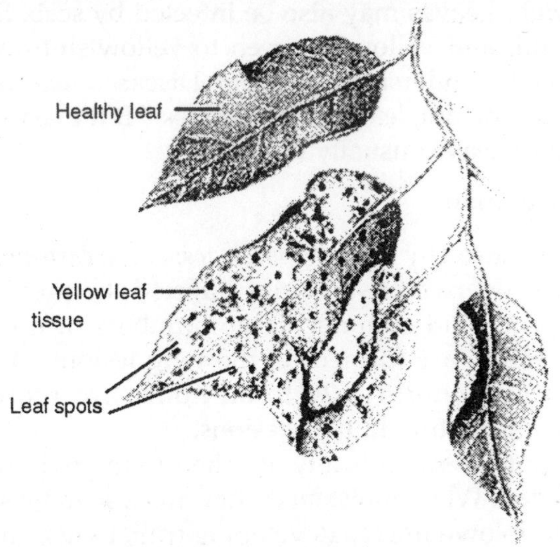

Fig. Cherry Leaf Spot Disease

The principle loss from scab is the unsightly spotting of the fruit skin, but premature defoliation and a dieback of infected twigs can also occur. Losses are generally greater on peaches than on nectarines, plums, and apricots. Wet conditions during spring and early summer after petal fall are necessary for a severe attack by the scab fungus. The disease is usually more serious in low-lying, moist, shady areas where air movement is poor.

Symptoms

Symptoms first appear on fruit when they are half formed to nearly fullgrown, about six to seven weeks after petal fall. Small, round, olive-green spots about 1/16-1/8 inch in diameter develop on the fruit. Spots are most common near the stem end on the side of the fruit exposed to the sun. Spots are superficial and slowly enlarge.

They may merge to form large, irregular blotches that turn velvety, dark olive-green or black. Severely infected fruit may be stunted, become misshapen, or crack open. When fruits crack open, they are often invaded by other fungi that rot the fruit. Scabby fruit may also drop prematurely and do not ship or store well. Leaves may also be infected by scab. Small (1/4 inch), round, and yellowish-green to yellowish-brown spots develop on the underside of the leaf. Diseased leaf tissue may dry up and drop out, leaving "shot-holes." If the season is wet, scab-infected leaves usually drop early.

Causal Organism

Scab is caused by the fungus *Cladosporium carpophilum*. The scab fungus overwinters in twigs infected the previous year. During spring and summer, large numbers of microscopic spores (conidia) are formed on the twig lesions. At 100 per cent relative humidity, 20 to 30 hours are required for abundant sporulation in twig lesions.

The spores remain firmly attached to the twig until they are moistened. When moistened, they are spread by splashing rain or windblown mist to developing fruit, twigs, and leaves. Spores are produced in the greatest numbers about two to six weeks after the shucksplit stage of development during warm, wet weather. Spore germination and fungus growth is most rapid at 65-75 degrees F.

If weather conditions are favourable, infection begins to occur at about shuckfall. The fruit remain susceptible until harvested. Forty to 70 days elapse from the time the spore lands on the fruit until the disease is visible. Thus, the disease is usually not observed until the fruit are well grown. Spores from the fruit reinfect the twigs and leaves, completing the disease cycle.

Control

- When planting an orchard, avoid low-lying or shaded sites with poor air circulation and soil drainage. Any practice that promotes faster drying of fruit and foliage will help reduce the risk of infection. Destroy nearby wild or neglected peaches, nectarines, plums, and apricots.
- Remove and destroy scab-blighted twigs before growth starts in early spring.
- Backyard growers should remember that infections are generally superficial. If infection is not severe, fruit quality is little affected. Peeling infected fruits make them quite acceptable for processing (canning or freezing) or eating fresh.

CHERRY LEAF SPOT

Cherry leaf spot is one of the most serious diseases of both sweet and sour cherries in the Midwest. The disease mainly affects the leaves, but lesions may also appear on fruit, petioles, and fruit stems (pedicels). Diseased leaves drop prematurely, and severely affected trees may be defoliated by mid-summer. Early and repeated defoliation can result in: 1) dwarfed and unevenly ripened fruit with poor taste; 2) devitalized trees that are more susceptible to winter injury; 3) death of fruit spurs; 4) a reduction of fruit set and size; 5) small and weak fruit buds; 6) reduced fruit growth; and 7) eventual death of the tree.

Symptoms

During the latter part of May and the first half of June,

small circular purple spots appear on the upper surface of the leaf. Spots gradually enlarge to about l/4 inch in diameter and turn reddish-brown. Lesions may emerge to produce large, irregular spots. Whitish-pink masses of sticky spores (conidia) form within the spots on the undersides of infected leaves during periods of damp weather. After six to eight weeks, the centres of the spots may dry up and drop out, giving a "shot-hole" appearance. The "shot-hole" effect is more common on sour than on sweet cherries.

The most conspicuous symptom, especially on sour cherries, is the golden yellowing of older infected leaves before they drop off. Although this symptom does not occur every season, the spotting of infected leaves is always visible. Spots similar to those on the leaves may also form on leaf petioles and fruit pedicels, causing fruit to ripen unevenly. Spots usually do not form on fruit.

Causal Organism

Cherry leaf spot is caused by the fungus, *Blumeriella jaapii* (previously called *Coccomyces hiemalis*). The fungus overwinters in dead leaves on the ground. In early spring (about petal fall), fungal fruiting bodies called apothecia develop in these leaves. Spores (ascospores) are produced in the apothecia and are forcibly discharged during rainy periods for about six to eight weeks, starting at petal fall. The optimal temperatures for ascospore discharge are 61 degrees F (16 degrees C) and higher. Very few ascospores are discharged at temperatures below 46 degrees F (8 degrees C).

These ascospores are spread by wind or splashing rain drops to the green, healthy leaves and serve as primary inoculum for disease. The ascospores stick to the leaf surface, germinate in a film of water, and within several hours at the proper temperature, penetrate the leaf through stomata (natural openings) on the underside of the leaf. The small purple spots soon appear on the upper surface. Incubation time, from fungus penetration to the appearance of the spots, varies with temperature.

Under damp conditions and with temperatures between

60 and 68 degrees F (15-20 degrees C), the period may be as short as five days. When rain and dews are absent and at lower temperatures, as long as 15 days may be required before symptoms appear.

Once lesions have developed, masses of secondary or summer spores (conidia) are formed from the slightly concave eruptions (acervuli) on the underside of the leaf. This mass of conidia provides the white appearance to the underleaf lesions. Conidia are spread to other leaves by splashing raindrops and are capable of causing new infections (each producing thousands of additional conidia) under the temperature and wetness conditions listed in Table.

Serious leaf spot damage usually occurs in years with numerous rainy periods throughout late spring and summer, when repeated secondary infection cycles allow the disease to snowball into an epidemic.

Control

- Collect and destroy the fallen cherry leaves in late autumn. The fungus overwinters in these leaves. This practice should be quite beneficial for backyard growers with one or a few trees, but is generally not considered practical for large commercial plantings.
- Select a planting sight that is always exposed to direct sunlight and has good air circulation and soil drainage. Proper pruning to open the canopy will increase sunlight penetration and air circulation. Any practice that promotes faster drying of leaves will reduce the risk of infection.
- Especially in commercial plantings, the disease is controlled primarily with fungicide sprays.

SOOTY BLOTCH AND FLY SPECK OF APPLE

Sooty blotch and fly speck are two different diseases. Both diseases are widespread throughout the Midwest, and commonly occur together on the same fruit. Sooty blotch and fly speck cause a discolouration or blemish of near-mature fruit. The discolouration is superficial, and while neither disease actually damages the fruit, the presence of disease reduces the grade and market value of the fruit. Although all apple varieties are susceptible to infection by both fungi, symptoms are most severe on yellow or light-coloured varieties such as Golden Delicious or Grimes. Both diseases are most common during years with a cool, wet spring, rains in late summer and low temperatures in early fall.

Symptoms

Brown to dull black, sooty blotches with an indefinite outline form on the fruit surface. Blotches may be 1/4 inch in diameter or larger. The blotches may coalesce to cover practically the entire fruit. The sooty blotch fungus is restricted to the outer surface of the fruit, and in many cases the blotches can be easily rubbed off. However, if infection occurs early in the season, you may need to rub or bleach the fruit vigorously to remove it.

Groups of 6 to 50 or more black and shiny round dots that resemble fly excreta appear on the surface of the fruit. The individual "fly specks" are clearly separated and can be easily distinguished from sooty blotch. Like sooty blotch, fly speck infections are superficial; however, they are usually harder to rub off than sooty blotch.

Causal Organism

The fungus Gloeodes pomigena causes sooty blotch, and Zygophiala jamaicensis causes fly speck. Both fungi overwinter on twigs of various wild woody plants, especially wild blackberry and raspberry canes. Both fungi require free water on the fruit surface to infect.

SOOTY BLOTCH

The "sooty blotch" or "smudge" appearance on affected fruit results from the presence of hundreds of minute, dark fungal fruiting bodies (pycnidia) that are interconnected by a mass of loose, interwoven dark hyphae (fungal filaments). In spring, pycnidia on wild plants produce large numbers of spores (conidia) that ooze out and collect in a gelatinous mass.

The conidia are then spread by water splash or wind blown mists into orchards from late May or early June until fall. The fungus first affects apple twigs, then secondary colonies are initiated on the fruit. Cool, humid weather is essential for disease development.

The disease does not develop when temperatures reach 85 degrees F. When May and June are cool and moist and are followed by a hot July and August, sooty blotch symptoms often do not appear for two to three months. The disease is absent or rare when hot, dry weather prevails until close to harvest time.

The disease is most severe when cool, rainy weather in the spring is coupled with late summer rains and low temperatures in early fall. Under ideal conditions, the incubation period from the time the fungus reaches the fruit to the appearance of symptoms may be as short as five days. In the orchard, the incubation period usually lasts three to four weeks on fruits that are 42 to 45 days old.

FLY SPECK

The individual "fly specks" are sexual fruiting bodies (ascocarps) of the fungus. Starting in late spring, the fungus produces spores on wild hosts. These spores are carried by wind into the orchard. When spores come into contact with the fruit under the proper environmental conditions, they germinate and infect. Symptoms can develop within 15 days under favourable environmental conditions (65 degrees F or 18 degrees C).

Control

- Select an orchard site that always has full sunlight, good air circulation, and good soil (water) drainage.
- Prune trees annually to an open centre for maximum air circulation. Both diseases are most prevalent in the damp, low, shaded areas of the orchard. Any practice that opens up the trees to greater air movement and promotes faster drying greatly aids in control.

- Remove or destroy nearby wild or neglected apple trees.
- Backyard growers should remember that the disease is superficial and rarely affects the quality of the fruit. Removal of the fungus by washing, rubbing, or peeling the fruit results in fruit that is acceptable for cooking or eating fresh.
- Especially in commercial plantings, fungicide sprays are important for controlling these diseases.

WHITE ROT AND BOTRYOSPHAERIA CANKER OF APPLE

White rot of apple fruits is also referred to as "Bot Rot" or Botryosphaeria rot. The fungus that causes fruit rot can also cause a canker on limbs and other above-ground woody portions of the tree. The canker phase of the disease is most severe in trees weakened by drought, winter injury, sunscald, poor pruning, low or unbalanced nutrition, and other plant diseases.

The fruit rot phase can be sporadic in appearance, being serious one season and difficult to find in the following season. The *Botryosphaeria* fungus attacks a wide range of woody plants that are common. Dutchess, Golden Delicious, Grimes Golden, Gallia Beauty, Rome, and Yellow Transparent apple varieties are all very susceptible to fruit rot. Jonathan and Red Delicious are less likely to be affected than other varieties.

Symptoms

Fruit

At first, small, reddish-brown spots appear around the lenticels. The spots enlarge and become slightly depressed. On yellow-skinned varieties, these spots may be bordered by one or more red "halo" rings. Spots on red-skinned apples often become bleached. The tissue under the spots is soft and egg shaped, with the long axis parallel to that of the core. Several spots may merge to involve all or much of the fruit. As the rot progresses, the skin colour becomes dark brown and superficially resembles black rot, except that with black rot the decayed tissue is firm, instead of soft and mushy. Beads of exudate appear on the surface of fruits completely rotted by white rot.

Small, black fruiting bodies (pycnidia) that are filled with spores (conidia) develop on the surface of rotted fruits during warm, moist conditions. Mature fruits are most susceptible to the disease. Apple fruits often become infected from midsummer on without showing external symptoms. Thus, fruit may be infected without symptoms appearing in the orchard. Disease development is checked when these apples are placed in cold storage. However, when they are removed from cold storage an internal rot develops and the fruit may deteriorate very rapidly at room temperature.

Twigs and Limbs

Small, circular spots or "blisters" appear on the twigs in June and July. The spots enlarge, become somewhat sunken, and fill with a watery fluid. The fungus may grow rapidly through the tissues to form slightly sunken, dark-coloured cankers that may extend to the cambium on very susceptible apple varieties. Under favourable conditions, several cankers may fuse to girdle and kill large limbs.

On older cankers, the outer bark becomes tan to orange and papery, and the margins of the canker crack and fissure. The outer bark sloughs off and the underlying tissue appears slimy. In the fall, twig and limb cankers stop growing and may split along the edges. Rings of small, black, spore-producing bodies (pycnidia and perithecia) are formed on the surface of the cankers or under the papery outer bark. The following spring, a canker may resume growth or be corked off and become inactive.

Casual Organism and Disease Cycle

Botryosphaeria canker and fruit rot (white rot) is caused by the fungus *Botryosphaeria dothidea*. The fungus overwinters as black pycnidia and perithecia in a wart-like stroma on living and dead cankered limbs and in rotted fruits. The fungus is also commonly found on fire-blighted twigs or cankers. Wounds or breaks in the epidermis are necessary for the fungus to penetrate. Spores (ascospores) are forcibly discharged from perithecia during spring rains.

Another type of spore (conidia) is produced within pycnidia and ooze out in tremendous numbers. They are then washed and rain-splashed to other parts of the tree throughout the summer.

Apple fruits may become infected fairly early in the season, but rotting does not develop much until the fruit is almost mature. At temperatures above 75 degrees F (24 degrees C), mature fruit may rot completely within a few days after infection. The development of Botryosphaeria canker and fruit rot is favoured by any condition that reduces tree vigour.

Control

Control of white rot is best achieved through an integrated programme of cultural practices and chemical control measures.

- Sanitation is critical for effective control. Piles of prunings are an important source of inoculum and should be removed from the perimeter of the orchard or burned. Prunings can be left on the orchard floor if they are chopped with a flail mower, which removes much of the bark and allows them to decompose faster.

  Removal of mummified apples and pruning out dead wood in the tree are important for reducing the inoculum within the tree. Pruning out current-season shoots infected with fire blight is also important, because they can be colonized and serve as an inoculum source during the same growing season.
- Any practice that helps to maintain trees in a healthy, vigorous condition is critical for controlling the canker phase of the disease.

  Cankers generally develop only on stressed or weakened trees. Prune trees annually and maintain a balanced fertility programme based on soil and

foliar nutrient analysis. Cankers generally develop rapidly on winter-injured trees.
- The use of fungicides combined with good sanitation is beneficial for controlling the fruit rot phase of the disease. Fungicides are not effective for controlling the canker phase of the disease on weakened trees.

RUST OF APPLE

Three rust diseases occur on apple and crabapple. All are caused by different species of the fungus *Gymnosporangium* and have various junipers and red cedars (*juniperus* species) as an alternate host. Apples are generally most susceptible to infection by the rust fungi during the period from early bloom until about 30 days after bloom.

Cedar-Apple Rust on Apple

On leaves, pale yellow spots appear on the upper surface during May or June. Spots are up to 1/4 inch in diameter, turn orange with time, and often have a reddish border. Small black fungal bodies (pycnia) form within the spots and may exude an orange fluid. In time, yellow spots develop on the underside of the leaf. These spots thicken, and during late spring and early summer a number of small, orange-yellow tubular projections (aecia) appear.

These develop into open, cylindrical tubes that split toward the base into narrow strips and curl backward. Infected leaves may turn yellow and drop. Defoliation of rusted leaves is most common in dry summers. On fruit, similar yellow-orange spots appear, usually at or near the calyx end.

These spots usually occur on immature fruit and are much larger than the spots on leaves (up to 3/4 inch in diameter). The light green colour of the young fruit becomes a darker green around the infected area. The tube-like aecia may form on the slightly raised fruit lesions. Infected fruits are often stunted and misshapen, and may drop early.

Cedar- Quince Rust on Apple

Cedar-quince rust only affects the fruit of apples. Infected fruit become puckered at the blossom end and later develop a sunken, dark green area. The flesh under the sunken, dark green area becomes brown and spongy.

The formation of pycnia and aecia on infected fruit is rare. Apples are susceptible to cedar-quince rust during the period from early bloom through third cover. Cedar-Hawthorn Rust on Apple Leaf spots similar to those caused by cedar-apple rust develop on apple and crabapple. Larger, gray to brown spots form on leaves of hawthorn. Few aecia form on apple and crabapple. Fruit infection on apple is rare. Defoliation and deformation of fruits and twigs may occur on hawthorns.

Causal Organisms

Cedar-apple rust is caused by the fungus Gymnosporangium juniperi-virginianae, cedar-quince rust by Gymnosporangium clavipes, and cedar-hawthorn rust by Gymnosporangium globosum. The disease cycle for all three rusts is essentially the same. The disease cycle for cedar-apple rust will be presented. The fungus overwinters as mycelium in galls in juniper.

Large yellow to orange gelatinous sporehorns are formed on the galls in the spring and spores (teliospores) are produced. Each teliospore germinates and produces four to eight sporidia or basidiosproes.

As sporehorns begin to dry, the sporidia are forcibly discharged into the air and carried by wind to nearby apple leaves, fruits, and twigs. About 30 days after apples have bloomed, the sporehorns have discharged all their spores and most apple leaves are no longer susceptible. Within five or six hours after landing on the leaf, sporidia become attached to the surface, germinate, and penetrate the cuticle and upper leaf surface.

After 10 to 14 days, the yellow spots develop on the upper leaf surface. The orange to black pycnia develop in the spots and several weeks later, the aecia form on the underleaf surface. The aecia produce another type of spore (aeciospores) that are carried by wind to junipers.

When the aeciospores contact a juniper twig, they become

firmly attached and germinate in warm moist weather of late summer or early fall and penetrate the twig. A young, pea-size, greenish-brown gall develops. The gall enlarges the following year, but does not produce sporehorns with teliospores until the second spring. The complete disease cycle requires almost two years.

Control

- Grow resistant or immune apples, crabapples, and junipers. When buying trees, check with the nursery about rust resistance.
- Destroy nearby, worthless or wild junipers infected with rust galls.
- Where rusts are a problem, follow a recommended fungicide spray programme.

BOTRYTIS BUNCH ROT OR GRAY MOLD OF GRAPE

Botrytis bunch rot is caused by the fungus Botrytis cinerea. This fungus is very common in nature, and causes diseases on a variety of unrelated crops. Bunch rot can cause serious losses on highly susceptible grape varieties. Although berries of all grape varieties are susceptible to bunch rot, losses generally are greater on tight-clustered varieties of Vitis vinifera and French Hybrids. Losses result from the rotting of berries in the field or in storage.

Symptoms

Infection of ripe berries is the most common and destructive phase of this disease. Infected berries first appear soft and watery. The berries of white cultivars become brown and shriveled, and those of purple cultivars develop a reddish colour. Under high relative humidity and moisture, infected berries usually become covered with a gray growth of fungus mycelium.

One or a few berries within the bunch or the entire bunch may be affected. Generally, healthy berries touching infected berries will become infected. Rotted berries generally shrivel with time and drop to the ground as hard mummies. The fungus also can cause a blossom blight that can result in significant crop loss early in the season. Although uncommon, leaf infections also occur, but appear to be of no economic importance. Leaf infection begins as dull, green spots, commonly surrounded by a vein. The spots rapidly become necrotic lesions.

Casual Organism and Disease Cycle

Botrytis bunch rot is caused by the fungus Botrytis cinerea. The fungus overwinters in grape mummies, dead grape tissues, and other organic debris in and around the vineyard, as well as on a multitude of alternate plant hosts. Because of its wide host range, growers always should assume that the fungus is present in the vineyard. In spring, the fungus germinates from small, dark, hard resting structures known as sclerotia. The fungus then produces spores (conidia) that spread the disease.

These spores are produced throughout the growing season. As blooms die, the spores germinate and colonize dead flower parts. Using the dead tissue as a food base, the fungus invades living tissue. After penetrating the berry, the fungus may remain dormant until the fruit sugar content increases and the acid content decreases to a level that supports fungus growth. Symptoms then develop readily under warm, moist conditions.

Berries that escape bloom-time infection may become infected at or near harvest under favourable environmental conditions. Any wound on the berry provides an excellent infection site for the fungus even in the absence of favourable environmental conditions.

Birds, insects, hail, and powdery mildew are common causes of wounds. Swelling during ripening in tightly packed clusters causes pressure that also can rupture the berries. Wet and humid conditions around the berries and leaves greatly enhance disease development. The longer wet conditions persist, the greater the probability of infection, even to undamaged berries.

Warmer temperatures also favour infection. At 54 to 75

degrees F, infection occurs in 12 to 24 hours, while at 37 degrees F, 60 to 72 hours are required.

Control

- Promote good air circulation and light penetration by proper pruning, controlling weeds and suckers, and positioning or removing shoots for uniform leaf development. Where possible, rows should be planted in the direction of the prevailing wind. Good air circulation and light penetration promote faster drying of plant parts and reduce the risk of disease. Removal of leaves around clusters on mid- or low-wire cordon-trained vines before bunch closing has been shown to reduce losses caused by Botrytis in New York and California vineyards, due to improved air circulation and improved spray penetration and coverage.
- Prevent wounding by controlling insects, birds, and other grape diseases.
- Growth regulators that lengthen the rachis and separate the berries in tight-clustered cultivars can reduce the damage from berries being crushed within the cluster; thus, reducing infection and spread of Botrytis.
- In commercial vineyards, effective fungicides applied at appropriate times during the growing season provide significant control.

VERTICILLIUM WILT OF RASPBERRY

Verticillium wilt is one of the most serious diseases of raspberry. This disease is caused by a soil-borne fungus and reduces raspberry yields by wilting, stunting, and eventually killing the fruiting cane or the entire plant. The disease is usually more severe in black than in red raspberries. Blackberries are also susceptible to the disease, but seldom suffer severe losses.

Verticillium wilt is usually a cool-weather disease and is most severe in poorly drained soils and following cold, wet springs. The appearance of symptoms on new canes frequently coincides with water stress caused by hot, dry mid-summer weather.

Symptoms

The symptoms usually appear on black raspberries in June or early July, and on red raspberries about a month later. The lower leaves of diseased plants may at first appear to have a dull green cast as compared to the bright green of normal leaves. Starting at the base of the cane and progressing upward, leaves wilt, turn yellow and drop. Eventually, the cane may be completely defoliated except for a few leaves at the top.

Black raspberry canes may exhibit a blue or purple streak from the soil line extending upward to varying heights. This streak is often not present or difficult to detect on red raspberries. The final effects of the disease are observed on fruiting canes that were infected the year before. In the spring, many of the diseased canes will be dead. Others will be poorly developed and have shriveled buds. The new leaves are usually yellow and stunted. Infected canes may die before fruit matures, resulting in withered, small, and tasteless fruit.

Casual Organism

Verticillium wilt is caused by the fungus, *Verticillium albo-atrum*. It is a very common soilborne fungus, and has been reported to cause wilt on more than 160 different kinds of plants including strawberries, eggplant, tomatoes, potatoes, stone fruits and peppers. The *Verticillium* fungus overwinters in the soil and plant debris as dormant mycelium or black, speck-sized bodies called microsclerotia. The fungus can survive in the soil for many years.

When conditions are favourable, the microsclerotia germinate and produce threadlike fungus filaments (hyphae). These hyphae can penetrate the root directly; but invasion is aided by breaks or wounds in the roots. Once inside the root, the fungus grows into the water-conducting tissue (xylem). The destruction of water-conducting tissues prevents the movement of water from the roots to the rest of the plant; thus, the plant eventually dies. The fungus produces microslerotia in infected tissues. When these tissues die and are returned to the soil, the disease cycle is completed.

Control

Applications of fungicides are ineffective in control. Soil fumigation has provided excellent control in some locations but is generally very expensive. Reintroduction of the pathogen into fumigated soils, accompanied by a rapid buildup in pathogen populations, is a major concern with using soil fumigation. Rotations (3- to 4-year) with nonsusceptible crops have been recommended for control in Canada but were not effective in California.

Only disease-free nursery stock from fields known to be free of *Verticillium* should be used to establish new plantings. Satisfactory resistance in commercial raspberry cultivars is not available. It is generally recommended that raspberries not be replanted in an area where the disease has been a problem. If they are replanted in an infested site, soil fumigation should be considered.

BLACK ROOT ROT OF STRAWBERRY

Black root rot is a serious and common problem of strawberries. The term "Black root rot" is the general name for several root disorders that produce similar symptoms. The disorders are not clearly understood and are generally referred to as a root-rot complex. For this reason, it is difficult to discuss black root rot as we do other diseases which usually have a specific cause. Black root rot has been found in every strawberry growing area of the United States, and a considerable incidence of black root rot has been observed in recent years.

Symptoms

Black root rot is most common in fields with a long history of strawberry production. Symptoms begin with some plants in a field showing reduced vigour, often in low or wet spots or in portions of the field where the soil has become compacted. This decline in vigour usually begins during the first fruiting year. The symptoms are most apparent the last couple of weeks before harvest.

Although severely affected plants may die before harvest, it is more common for diseased plants to continue living but become stunted and produce a reduced crop of small berries. The percentage of plants affected in any individual field usually increases significantly the year following the first appearance of symptoms.

Diagnosis is made by digging up declining plants and examining their root systems, about the time that fruit begin to colour.

Abundant fleshy white roots and fine lateral roots will be seen on healthy plants, and the interior of the older woody roots is yellowish - white. With black root rot, there is usually a loss of many fine lateral roots, and irregular black patches occur along the length of the fleshy white roots. In severely affected plants, these black patches grow together so that no white roots are visible. The interior of infected older woody roots turns black.

Potential Causes

Several different fungi have been implicated as causes of black root rot, as have certain environmental stresses such as cold injury, soil compaction, and excessive water in the root zone.

In some soils, black root rot has been associated with an interaction between a particular soilborne fungus and the lesion nematode *Pratylenchus penetrans*. It is likely that black root rot symptoms result from one or more of the following:

- Gradual buildup in the soil of disease— causing microorganisms and nematodes when strawberries are grown with inadequate rotation;
- Interaction of these organisms with environmental or other stress factors such as herbicide injury, winter or cold injury, and excessive soil moisture that might make plants more susceptible to attack; and

- Certain soil conditions such as heavy (clay) or poorly drained soils that might favour the activity of disease - causing fungi and/or inhibit the ability of the strawberry plant to produce new roots to compensate for their damage. Additional factors may also be involved.

Control

Because several factors appear to be involved in the black root - rot complex, no general control measure is totally effective. The following may help to reduce its incidence:

- Always start plantings with healthy white-rooted plants from a reputable nursery.
- Rotate out of strawberries for at least 2-3 years before replanting.
- Minimize soil compaction and increase tilth by incorporating organic matter, such as straw from a rotational grain crop.
- Avoid heavy, wet soils and improve drainage in marginal soils by tiling or planting on raised beds.
- Preplant fumigation of the soil is sometimes helpful, but not always.

PHYTOPHTHORA ROOT AND CROWN ROT OF FRUIT TREES

Phytophthora root and crown rots (sometimes called collar rot) are common and destructive diseases of fruit trees throughout the world. Apple, cherry, and peach trees are usually attacked. Pear and plum trees appear to be relatively resistant. Trees declining and dying from Phytophthora root and crown rots are frequently misdiagnosed as suffering from "wet feet" (root asphyxiation) or are sometimes confused with those suffering from winter injury.

Symptoms

Diseased trees are commonly found in poorly drained areas of the orchard or yard. Heavy, wet soils that remain saturated for extended periods of time are required for disease development. Above-ground symptoms vary between tree species, but generally include reduced tree vigour and growth, yellowing or chlorosis of leaves, and eventual collapse or death of the tree. Infected trees may decline slowly over one or more years, or they may collapse and die rapidly after resuming growth in the spring.

Trees may also appear healthy in the spring, but die suddenly in the latter part of the growing season. Rapid death of trees usually occurs following excessively wet periods. On trees that decline gradually, a reddish or purple discolouration of the leaves often occurs in autumn, while leaves on healthy trees remain green. To observe below-ground symptoms, you need to remove several inches of soil around the base of the declining tree.

A diagnostic reddish-brown discolouration of the inner bark and wood can be observed after cutting away the outer bark layer. A sharp line demarcates the reddish-brown (diseased) and white (healthy) portion of the crown. Similar symptoms can be found on roots, but it is generally difficult to see root symptoms without removing the tree.

This reddish discolouration and line of demarcation between diseased and healthy tissue distinguishes Phytophthora root and crown rot from other causes of tree decline and collapse such as "wet feet" (drowning) or winter injury. Roots on trees killed by excessive water are usually completely black (have no line of demarcation) and often times have an unpleasant smell.

Discolouration from winter injury is usually confined to the above-ground part of the trunk, particularly on the southwest side of the tree, while the below-ground portion of the tree may still appear healthy.

Causal Organism and Disease Cycle

Phytophthora root and crown rots are caused by several *Phytophthora* species. These are all soilborne fungi, many of which are common inhabitants of most orchard soils. Some species that are not common inhabitants may be introduced to the orchard on contaminated planting stock or through movement of contaminated soil. While some species are much more destructive than others, depending on the type of fruit tree and rootstock, all species require extremely wet or saturated soils in order to infect and cause significant damage.

These fungi overwinter and persist in soil as mycelium in infected wood or as thick-walled spores (oospores). Oospores remain viable in the soil for long periods of time (years). When soils are wet, oospores germinate to form thread-like fungal filaments (mycelia).

Mycelia from germinated oospores or from infected tissues produce reproductive structures called a sporangia. These sporangia are filled with infective spores called zoospores. Zoospores are released from sporangia only when soil is completely saturated with water (standing water).

The zoospores use flagella to swim to susceptible plant tissue where they infect. They may also swim to the soil surface and move over longer distances in runoff water. The longer the period or periods of soil saturation, the greater the risk of infection. Some rootstocks appear to be most susceptible during spring and autumn, which are also the times of year when soil temperatures are most conducive to fungus growth and zoospore production. Rootstock susceptibility and fungus activity are both low in the winter when trees are dormant.

Control

Control of Phytophthora crown and root rots is most successful using an integrated programme of cultural practices, choosing the most resistant tree species or rootstock, and when necessary, chemical control.

- Avoid sites that drain slowly or poorly or are subject to periodic flooding. Marginal sites should be modified (install drain tiles, create diversion ditches, rip underlying pan layers) to provide the additional drainage recommended for growing tree fruit crops. Planting trees on ridges or berms will raise their crowns above the primary zone of zoospore activity and provide an important margin of safety, especially in a wet year.

- Select rootstocks or tree species that are less susceptible to Phytophthora and are best adapted to your individual site. Pears are the most resistant tree fruit crop and are most likely to remain healthy in a relatively wet site. Among apple rootstocks, seedlings are relatively resistant.
  Among dwarfing-apple rootstocks, M-9, M-2, and M-4 are relatively resistant. The Canadian rootstock Ottawa-3 has M-9 type resistance. M-7 and MM-111 are moderately susceptible; M-26 and MM-106 are susceptible; and MM-104 is highly susceptible.
  Among stone fruits, plums are relatively resistant, whereas the remainder are susceptible to very susceptible. Mahaleb is the most susceptible cherry rootstock, whereas Mazzard, Morello, and Colt are somewhat more resistant and would be recommended on heavier soils
- Soil fumigation is generally considered ineffective because it never completely eradicates the fungus from orchard soils, and the Phytophthora fungi are easily reintroduced into fumigated soil.
- New fungicides have recently been developed, which are effective in controlling these diseases when used preventively, but they are seldom effective in reviving trees once the crown has become infected and moderate symptoms of decline have appeared. Fungicides are most effective when used in combination with the cultural practices described above.

BLACK ROT AND FROGEYE LEAF SPOT OF APPLE

Black rot and frogeye leaf spot are phases of a widespread and damaging disease of apple and crabapple. The fruit rot phase is called black rot and on the leaf it is called frogeye leaf spot. The disease can result in losses from

- A rotting of fruit before harvest and in storage.
- A weakening of the tree from defoliation, and
- A blighting and dieback of twigs and limbs caused by girdling cankers. The premature dropping of

infected leaves can result in small, poor-quality fruit and reduces crop yield the following year. All apple varieties appear to be equally susceptible to fruit rot. Jonathan and Winesap appear to have the greatest susceptibility to leaf infection.

Symptoms

Fruit

The disease usually starts at the calyx end of the fruit. The fungus usually enters the fruit through wounds caused by insects, hail, growth cracks, or an open calyx tube. At first, a light brown spot forms on the fruit. Usually only one spot occurs per fruit. With time, the spots enlarge and commonly develop a series of brown and black concentric bands or rings.

The rotted fruit finally turns black. The decayed tissue remains firm to leathery, and holds its original shape until the entire fruit is rotted. The completely decayed fruit finally dries and shrivels into a wrinkled black "mummy" which may remain on the tree a year or longer. Black, pimple-like fruiting bodies (pycnidia) of the causal fungus appear on the surface of rotted fruit. In cold storage, the flesh of black rot-infected fruits remains firm, in contrast to several other apple rots.

Leaves

Starting at petal-fall or somewhat later, small, purple specks appear on infected leaves. These specks enlarge to form spots 1/8 to 1/4 inch in diameter. The round to irregularly lobed spots develop a light brown-to-gray centre surrounded by one or more dark-brown concentric rings and a purple margin giving it a "frogeye" appearance. Black pycnidia, like those that appear on rotted fruit, may develop on the upper surface in the centres of the older leaf spots. These pycnidia help to distinguish frogeye leaf spots from similar spots caused by spray injury.

Twigs, Limbs, and Trunks

Small, slightly sunken, reddish-brown areas develop in

the bark. These areas slowly enlarge and darken to form cankers. Cankers may continue to expand a little each year, and may extend down the limb for 3 feet or more.

These areas remain somewhat sunken, except for the slightly raised and lobed margin. Cankers may appear as a superficial roughening of the bark; or the bark may be killed and conspicuously cracked, especially at the margins. In recently killed areas, the bark is firmly attached to the wood; but after a year or so, it cracks and falls away and can be easily removed from the wood. Black pimple-like pycnidia and another very similar fungal fruiting structure (perithecium) are usually abundant in older cankers.

Causal Organism and Disease Cycle

Black rot and frogeye leaf spot are caused by the fungus, *Physalospora obtusa*. The fungus overwinters in cankers, mummified fruits, and the bark of dead wood. In the spring, the black fungal fruiting bodies (pycnidia and perithecia) release conidia and ascospores, respectively. These two types of spores spread the disease to healthy leaves, fruit, and wood. The heaviest discharge of spores occurs around blossom time, but the production of conidia may continue during wet periods throughout the summer. The conidia can remain viable for at least one year. Leaf infection usually occurs during the petal-fall period.

Conidia become attached to the leaf and may germinate in a film of moisture within 5 or 6 hours. After germination, the fungus penetrates the leaf through natural openings in the under surface or through insect, hail or other wounds. Spore germination and infection are most rapid at 75 to 80 degrees F. Fruit infection can occur as early as petal fall; however, symptoms are usually not visible until mid to late-summer as the apple approaches maturity.

Control

Control of black rot is best achieved through an integrated programme of cultural practices and chemical control measures.

Sanitation is critical for effective control. Piles of prunings are an important source of inoculum and should be removed from the perimeter of the orchard or burned. Prunings can be left on the orchard floor if they are chopped with a flail mower, which removes much of the bark and allows them to decompose faster.

Removal of mummified apples and pruning out dead wood in the tree are important for reducing the inoculum within the tree. Pruning out current-season shoots infected with fire blight is also important, because they can be colonized and serve as an inoculum source during the same growing season.

Any practice that helps to maintain trees in a healthy vigorous condition is critical for controlling the canker phase of the disease. Cankers generally develop only on stressed or weakened trees. Prune trees annually and maintain a balanced fertility programme based on soil and foliar nutrient analysis. Cankers generally develop rapidly on winter-injured trees.

The use of fungicides combined with good sanitation is beneficial for controlling the fruit rot phase of the disease. Fungicides are not effective for controlling the canker phase of the disease on weakened trees.

PHOMOPSIS CANE AND LEAF SPOT OF GRAPE

For many years, the Eastern grape industry recognized a disease called "dead-arm," which was thought to be caused by the fungus *Phomopsis viticola*. In 1976, researchers demonstrated that the dead-arm disease was actually two different diseases that often occur simultaneously. Phomopsis cane and leaf spot (caused by the fungus *Phomopsis viticola*) is the new name for the cane- and leaf-spotting phase of what was once known as dead-arm. Eutypa dieback (caused by the fungus *Eutypa armeniacae*) is the new name for the canker- and shoot-dieback phase of what was once known as dead-arm.

We now propose that the name dead-arm be dropped. Growers should remember that Phomopsis cane and leaf spot and Eutypa dieback are distinctly different diseases, and their control recommendations vary greatly.

Disease incidence of Phomopsis cane and leaf spot appears to be increasing in many vineyards throughout the Midwest. Crop losses up to 30 per cent have been reported in some vineyards in growing seasons with weather conducive to disease development. Phomopsis cane and leaf spot can affect most parts of the grapevine, including canes, leaves, rachises (cluster stems), flowers, tendrils, and berries and can cause vineyard losses by:

- Weakening canes, which makes them more susceptible to winter injury.
- Damaging leaves, which reduces photosynthesis.
- Infecting cluster stems, which can result in poor fruit development and premature fruit drop.
- Infecting berries resulting in a fruit not near harvest.

Spots or lesions on shoots and leaves are common symptoms of the disease. Small, black spots on the internodes at the base of developing shoots are probably the most common disease symptom. These spots are usually found on the first three to four basal internodes. The spots may develop into elliptical lesions that may grow together to form irregular, black, crusty areas.

Under severe conditions, shoots may split and form longitudinal cracks. Although cane lesions often appear to result in little damage to the vines, it is important to remember that these lesions are the primary source of overwintering inoculum for the next growing season.

Leaf infections first appear as small, light-green spots with irregular, occasionally star-shaped, margins. Usually only the lower one to four leaves on a shoot are affected. In time, the spots become larger, turn black, and have a yellow margin. Leaves become distorted and die if large numbers of lesions develop. Infections of leaf petioles may cause leaves to turn yellow and fall off.

All parts of the grape cluster (berries and rachises or cluster stems) are susceptible to infection throughout the growing season; however, most infections appear to occur early in the growing season. Lesions developing on the first one or two cluster stems (rachises) on a shoot may result in premature withering of the cluster stem. Infected clusters that survive until harvest often produce infected or poor-quality fruit.

If not controlled early in the growing season, berry infection can result in serious yield loss under the proper environmental conditions. Berry infections first appear close to harvest as infected berries develop a light-brown colour. Black, spore-producing structures of the fungus (pycnidia) then break through the berry skin, and the berry soon shrivels. At this advanced stage, Phomopsis cane and leaf spot can be easily mistaken for black rot.

Growers should remember that the black rot fungus only infects green berries and will not infect berries after they start to mature. Berries become resistant to black rot infection by three to four weeks after bloom. Fruit rot symptoms caused by *Phomopsis* generally do not appear until close to harvest on mature fruit. Severe fruit rot has been observed in several vineyards.

Research has shown that berry infection can occur throughout the growing season; however, most fruit rot infections probably occur early in the season (pre-bloom to two to four weeks after bloom). Once inside green tissues of the berry, the fungus becomes inactive (latent), and the disease does not continue to develop. Infected berries remain without symptoms until late in the season when the fruit matures. Thus, fruit rot that develops at harvest may be due to infections that occurred during bloom.

Casual Organism and Disease Cycle

The fungus overwinters in lesions or spots on old canes infected during previous seasons and requires cool, wet weather for spore release and infection. The fungus produces flask-shaped fruiting bodies called pycnidia in the old diseased wood. These pycnidia release spores in early spring and are spread by splashing rain droplets to developing shoots, leaves, and clusters. In the presence of free water, the spores germinate and cause infection. Shoot infection is most likely during the period from bud break until shoots are six to eight inches long.

The optimum temperature for leaf and cane infections is between 60 and 68°F, and a wetness duration of at least six hours is required at these temperatures.

As the wetness duration increases, the opportunity for infection greatly increases. Lesions on leaves appear at seven to 10 days after infection. Fully expanded leaves become resistant to infection. Lesions on canes require three to four weeks to develop. The fungus does not appear to be active during the warm summer months, but it can become active during cool, wet weather later in the growing season. Pycnidia eventually develop in cane lesions and will provide the initial inoculum for infections during the next growing season.

Control

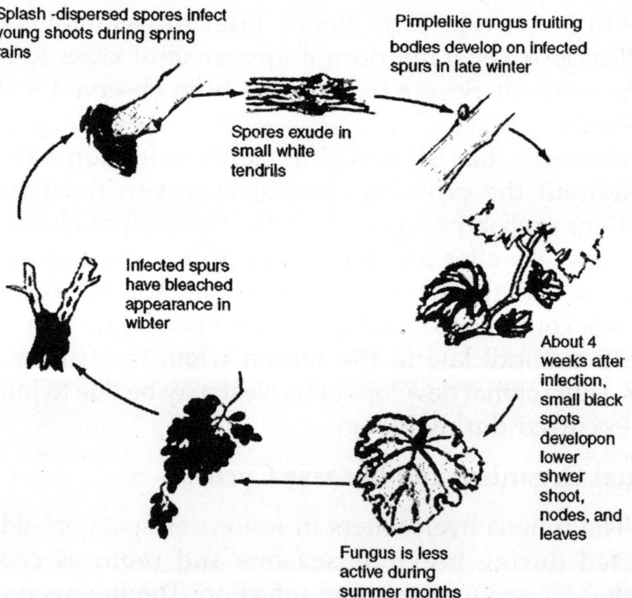

Fig. Phomopsis Cane and Leaf Spot.

- Select planting sites with direct, all-day sunlight (avoid shade). Good soil drainage and air circulation are also very important. Orient rows to take full advantage of sunlight and wind movement. Cultural practices that increase air circulation and light penetration in the vineyard will reduce wetting periods and should be beneficial for control.

- While dormant pruning, cut out infected canes and destroy them. Select only strong, healthy canes that are uniform in colour to produce the next season's crop.
- Proper timing of early-season fungicide sprays is important for control in commercial vineyards.

BITTER ROT OF GRAPE

Bitter rot has been recognized as a disease of grapes since 1887. The name was derived from the bitter taste that develops in infected berries. If 10 per cent of the berries used to make wine are infected with bitter rot, the wine may be undrinkable. The disease is more common in the southern grape growing regions of the United States, and could become a serious problem in southern region.

Bitter rot is easily confused with black rot, which is a common disease of grapes throughout. The ability to distinguish between these two diseases could be vital for growers and especially winemakers.

Symptoms

Bitter rot can affect young shoots, stems of fruit clusters, pedicels and fruit. If cluster stems are infected and killed early in the season, berries do not develop and remain attached to the stem. When cluster stems are infected late in the season, berries may fall off the bunch. This "shelling" of berries may be an important sign that bitter rot is occurring. The most obvious symptoms of bitter rot are on berries. The first symptom is a brownish, water-soaked lesion that develops on maturing fruit.

Bitter rot is a ripe rot of grape, and unlike black rot, does not infect green berries. This lesion spreads rapidly (often in concentric rings) and within a short period of time the entire fruit is affected. These infected berries usually maintain their shape and are dull brown in colour. In 2 to 3 days the skin of the berry is ruptured (more or less uniformly) by black pustules (fungal structures). Under moist conditions these pustules may coalesce to form irregular blisters over the fruit surface. The fruit skin is often torn by this point and the fruit shrivels in to black mummies closely resembling those of black rot. If berries become infected when they are overripe, the entire berry may become infected and fall off before the pustules form.

These infected but temporarily symptomless berries are usually the most bitter, and unfortunately, the most likely to be harvested and pressed (along with healthy grapes) into wine. It is a common practice to leave wine grapes on the vine past maturity in order to increase sugar content. This could become a dangerous practice if bitter rot becomes established in the vineyard.

Causal Organism

Bitter rot of grapes is caused by the fungus *Melanconium fuligineum*. The fungus overwinters in the vineyard on almost any plant debris, especially berry mummies. Spores (conidia) are produced from cushions of fungus tissue (acervuli) growing on plant litter. Fungus growth and spore production are favoured by warm, humid and rainy weather. Spores are spread by splashing rain.

Symptoms develop on berries one week after spores reach a healthy berry and in less time if the berry is wounded. The fungus colonizes the entire berry, including the seeds. Acervuli develop in the fruit skin, rupturing it as they mature. Conidia in the acervuli are spread to other plant parts by rain. Bitter rot increases in severity the farther south grapes are grown.

Control

Promote good air circulation and light penetration by controlling weeds and suckers, proper pruning, and positioning or removing shoots for uniform leaf development. Where possible, rows should be planted in the direction of the prevailing wind. Good air circulation and light penetration will promote faster drying of plant parts and reduce disease incidence. Prevent wounding of berries by controlling insects, birds and other grape diseases.It is important that growers and winemakers be able to distinguish between bitter rot and black rot. Because black rot infects only green berries, the fungicide sprays for controlling it are generally stopped when berries begin to change colour.

Stopping sprays at this time could be disastrous if bitter rot is present because it infects fruit only after the berries have changed colour. The table below is a list of characteristics that should be helpful in distinguishing between the two diseases.The fungicide spray programme that is recommended for control of the more common grape diseases should be beneficial in controlling bitter rot. Specific sprays for controlling bitter rot are generally not required.

CYTOSPORA CANKER OF SPRUCE

The most important disease of spruce in landscape plantings is Cytospora canker. This disease is caused by a fungus and is frequently found on Norway spruce and Colorado blue spruce and its cultivars. White spruce is also susceptible and there are a few reports on Serbian spruce. In addition to the spruces, Cytospora canker is sometimes found associated with Douglas fir, hemlocks, larches, and balsam fir.

Symptoms

Dying of a lower branch with subsequent needle browning is usually the first symptom. The brown needles may remain on the branches or they may fall off. As the disease progresses over several years, higher branches show damage. The actual cankers are often first seen at the base of branches near the main trunk of the tree. On the more susceptible species (Norway spruce), trunk cankers develop which may result in girdling and death of the tree. The bark of the cankered areas is not visibly different in colour, nor does it become sunken as in cankers on many deciduous trees. However, resin flow is usually associated with Cytospora canker and the white patches of dried resin are quite conspicuous on the bark. Resin flow can, however, be associated with any injury to branch tissue. Cankers often cannot be located without cutting into the bark. Removal of the outer bark from cankers reveals brown, dead areas of the inner bark and cambium. Within the cankered areas, a careful search using a magnifying hand lens will often reveal black pinhead sized structures that produce the spores of the pathogen.

Careful removal of layers of bark make these structures even more visible. During wet weather, yellow to orange coloured masses of spores oozing out of these black structures in tiny tendrils can be observed. Sometimes these tendrils or gelatinous spore masses are visible to the unaided eye.

CAUSAL FUNGUS

Cytospora canker of spruce is caused by the fungus, Cytospora kunzei var. piceae (also reported as Valsa kunzei var. piceae). The spores (conidia) described above are readily disseminated by splashing water, wind-driven rain, by man during pruning, and also very likely by insects and birds. The fungus generally becomes established through wounds.

Symptoms

Cytospora canker is more common on trees over 15 years old. This disease is more prevalent on trees of low vigour. Those trees with shallow roots, weakened by drought, low fertility, mechanical injury, or insect damage; and trees growing in an unfavourable site are more susceptible to Cytospora canker. The following practices lessen the likelihood of this disease.

- Avoid bark and stem injuries.
- Control insects and mites; especially spruce gall adelgids and spider mites.
- Fertilize according to horticulturists' recommendations.
- Water during extended dry periods. Water thoroughly so that soil is moistened 18 to 24 inches deep. A root irrigator may be needed to accomplish this.
- Follow accepted pruning practices.
- Vertically mulch to relieve soil compaction, poor aeration, and inadequate water penetration.

Once established, the following may aid in suppressing disease development. Remember that affected branches cannot be saved. Prune and remove or destroy affected branches. To lessen the spread of the fungus, prune only when the trees are dry. Pruning tools should be disinfested with 70 per cent alcohol between cuts. It will generally be necessary to prune back to the main trunk. No effective chemical control measures are available.

YELLOWING, DIEBACK AND DEATH OF NARROW-LEAFED EVERGREENS

Each year various narrow-leafed evergreens such as pines, Taxus and spruce are affected with needle yellowing and browning, dieback, poor vigour or death. These are problems often associated with one or more environmental stress factors. An explanation of some of these stress factors follows.

Wet Soil

Excessive amounts of water can result in a saturated soil, reducing oxygen levels to a point where small roots weaken or die. This root decline can be sudden or gradual and the roots may be invaded by various soil-borne fungi. Continuous wet conditions lead to progressively worse situations. If the top of the plant is unable to obtain the necessary water and nutrients, it declines or dies.

However, the evidence of death (needle browning) often occurs at a much later date. In fall, winter and spring, water accumulates and literally drowns the roots. The tops of the plant may not succumb until the following spring or summer, when hot weather first arrives and stresses the plant. Some evergreens appear to lose vigour and die back after 15 to 20 years. This often is the result of injury to the root system from moisture stresses.

Heavy soils may limit development of the root system; root damage easily upsets the top-to-root ratio. Dieback and poor growth are often evident. Changes in sub-soil drainage caused by construction will often cause roots on older plants to die back.

Drought or Dry Soil

The lack of water for long periods may result in symptoms similar to those caused by excess water. Clay soils often pull away from the roots as they dry, drying or breaking the fine roots. Drought stress may be especially noticeable in the summer on evergreens planted on well drained sites (sand or gravel), or where roots are in the top layers of heavy compacted soil. Excessive needle drop and poor vigour are often evident as a result of drought stress.

Winter Damage

Evergreen plants transpire or lose water from leaves during winter. If the soil moisture is low or roots are unhealthy, moisture in the needles is not replenished and needles are killed. Again, visible symptoms often do not appear until spring or early summer. Damage often appears on one side or on one branch of the plant, usually the side facing prevailing winds. Needles may turn brown one half or one third of the way from the tips. The extent of browning will be similar on all the needles on the branch.

De-icing Salt Damage

Injury from exposure to de-icing salt can occur on plants. Salt sprayed by traffic on wet roads can cause browned foliage, usually on the side nearest the road. Salt solution runoff also can injure plant roots. Entire plants may die. Needle yellowing and browning often begins at the tips and gets progressively worse. Sometimes soil tests conducted in late winter indicate high salts and confirm the diagnosis.

Herbicide Damage

Injury to evergreens by herbicides is difficult to assess. Symptoms are not always pronounced. Needle distortion may be slight, but root damage could be enough to limit water uptake. Tip damage to new growth is a common symptom with some herbicides. Look for needle distortion and twisting, or needle yellowing or browning, depending on the type of herbicide. On spruce, needle purpling and drop is common.

Air Pollution Damage

When atmospheric conditions allow buildup of smog or air pollutants, narrow yellowed bands may develop on the needles of susceptible plants. In other cases, the tips of the needles may turn brown. Ozone injury to white pine will cause a severe reduction in growth. If this continues for years, it is considered chlorotic dwarf disease. Trees under drought stress may be more prone to damage by air pollutants.

Low Light Needle Drop

Most narrow-leafed evergreens need full sunlight. Low light conditions may result in a slow decline of some evergreens such as junipers or arborvitae. An early symptom is foliage drop in the centre of the plant. The condition is common on plants existing in overgrown, old landscapes. Sometimes two plants will grow together. Both will begin to decline. In other cases, a deciduous plant nearby may begin casting a shadow on the evergreen plant during the morning or evening hours.

Transplant or Establishment Problems

Improper planting and poor after-transplant care may result in plant decline several years following transplanting. Common problems associated with planting and establishment include: burlap, especially synthetic burlap, left intact around the root ball; strings or wires left around the trunk; planting a containerized plant without disturbing the root mass; inadequate or inappropriate watering following transplanting; support wires left on the tree too long; setting the tree or shrub deeper than originally grown; and settling following transplanting. These problems are difficult to correct after symptoms have become apparent.

It is difficult to develop a control programme for environmental stresses. A wet soil suggests need for a better drainage system or that less watering is required. A dry soil indicates the need for a better and more uniform watering programme or using an effective mulch. Mulching, proper watering, use of a wind barrier or spraying with an antidesiccant in late fall helps lessen winter damage. Vertical mulching or core aerating will improve landscape soils. It will hasten drainage of excessive water, preserve necessary aeration during wet periods, allow sub-soil water penetration during dry periods and promote the formation of fine feeder roots.

Drill 1 or 2-inch wide, 18-inch deep holes in the soil on 12 to 20 inch centres under affected trees. Fill holes with a mixture of equal parts of peat and a coarse aggregate, such as pumice or calcined (baked) clay particles. Control de-icing salt damage by re-directing runoff water, installing splash or spray guards or by using tolerant plants in injury prone locations.

USING FUNGICIDE SPRAYS EFFECTIVELY

Fungicides can be an important component of the disease management programme. However, it is important to remember that their use should be integrated with the use of sound cultural practices, a knowledge of pathogen and disease biology, and disease resistance whenever possible.

Fungicides are only effective when infectious plant diseases that are caused by fungi are truly the cause of the problem. In many cases, pests and diseases follow other environmental imbalances and may not be the major problem. In cases such as these, a fungicide may help but is often not the total answer. Also, it is important to remember that fungicides are only effective if several rules are followed. First, the correct material must be selected.

This depends on correct diagnosis and identification of the pathogen. Second, the chemical must be applied at the right time of year and frequently enough to protect plant material adequately. Third, fungicides must be applied properly over plant surfaces. These three rules depend on making correct decisions based on correct knowledge. Too many people simply "spray and pray," and are often disappointed with the results.

Correct Diagnosis

One must be sure of what the problem is before

proceeding. The most effective fungicides in use today have been developed for specific situations and specific diseases. To use these chemicals, you must spend time making a correct diagnosis. Your state Extension specialists, plant disease diagnostic clinic, and county Extension agents can assist you with the proper diagnosis.

Selecting the Proper Material

Diagnosis leads to selection of the right material to do the job. Usually, several materials are effective against the type of disease you are dealing with. For instance, triforine, sulfur or triadimefon all control powdery mildews. Before selecting any chemical, read the label. Can you carry out the instructions? Is the plant type listed on the label? If so, the chemical is registered for use on the plant, and should be effective in providing disease control if used properly. If not, it is illegal to use that particular pesticide.

Use the Correct Method of Application

Foliar application fungicide sprays usually work in controlling infectious diseases because they act as a chemical barrier on leaf, stem or flower surfaces. When the pathogen arrives on the plant surface, it encounters this barrier and is unable to infect the plant. Effective fungicide use requires that this barrier be as complete as possible.

A spray method must provide the best combination of practical usefulness and good coverage. For many diseases, special attention must be given to undersurfaces of leaves, especially on the lower leaves of the plant.

The completeness of the barrier depends on how well the spray spreads and sticks to the plant surfaces. For this reason, spreader-stickers or spray adjuvants can be added to many sprays. Sometimes the product label alerts the user to these problems.

However, observing the spray deposit after you have finished some of the job may be the best way to decide if an adjuvant should be used. Hairy or waxy foliage is especially difficult to cover properly without a spreader-sticker.

Proper Timing of Fungicide Applications

Timing refers to when and how often the spray must be applied to effectively control a disease. The first application usually is made at a time close to but before the pathogen arrives on the plant surface.

This information is often provided on the pesticide label or is available in Extension literature from your local county Extension agent. In most situations, fungicides are not effective in controlling the disease if the pathogen has already entered (infected) the plant tissues. In many cases, specific information about the disease cycle may be needed to time the first application correctly. After the first application is made, the pesticide barrier is established on the plant surfaces. Effective use involves keeping this barrier active and complete throughout the time that the pathogen can arrive on and infect the plant. Modern fungicides are developed so that they do not persist in the environment for long periods of time. Rainwater, sunlight, microbial action and oxidation decrease effectiveness of the fungicide. Reapplication of the spray is needed in many cases to keep the fungicide barrier active.

Plant growth also affects the completeness of the barrier. As new leaves and shoots appear, they are unprotected and may be subject to infection. If so, they must be recovered with the barrier. The fungicide label gives reapplication guidelines, usually in ranges of 7-14 day intervals. If excessive rainfall or rapid growth of the plant occurs, the shorter interval between sprays should be used. If not, use the longer interval.

LEAF BLIGHT OF HAWTHORN

Leaf blight is a serious disease of a few very susceptible English hawthorn cultivars. Most other hawthorn species and cultivars are resistant or are not seriously affected.

Symptoms

The disease is first evident as small, angular, reddish-brown spots on the leaves. The spots have irregular margins and coalescence of spots often occurs resulting in larger irregular diseased areas. Diseased leaves yellow and fall prematurely. Sometimes, spots may be more prevalent near the margins of the leaves. When conditions are favourable for disease development, extensive defoliation occurs. It is usually mid-summer or after before the disease becomes prevalent. Small, black, raised dots develop in the centre of the spots. These are the spore masses of the causal agent. They are especially evident when the leaves are wet.

Causal Fungus

Leaf blight of hawthorn is caused by the fungus, Entomosporium thuemenii (Diplocarpon maculatum). The fungus survives from one year to the next in fallen diseased leaves and in inconspicuous stem spots. During May and June, spores are produced on these overwintered leaves. These spores are spread by splashing rain and initiate the disease on the current season's foliage. As the leaf spots develop, new spores are formed on the spots and rapidly spread the disease. Wet weather is favourable for rapid development because splashing water carries spores, and persistent water drops favour spore germination and infection.

Control

Plant Resistant Varieties

Washington hawthorn types are resistant to this disease. English hawthorn types are susceptible and are commonly infected. Although Washington types are resistant to leaf blight they are susceptible to rust diseases. If hawthorn rust is a serious problem in your area, it may be best to select plants other than hawthorns for landscape plantings.Since the fungus overwinters in the fallen diseased leaves, raking and destroying all leaves will help to manage leaf blight, but will probably not result in complete control.

Protection with Fungicides

Leaf blight can be prevented by spraying with fungicides. Control through use of fungicides depends on proper timing of the sprays. Based on presently available information, spraying at 10 to 14 day intervals from bud break through early July has given good control. Additional applications may be necessary during rainy seasons or when good control was not achieved with the earlier sprays. Fungicides registered for use on leaf blight of hawthorns include: thiophanate-methyl (Cleary's 3336, Domain, Fungo Flo); chlorothalonil (Daconil*); and mancozeb (Fore, Dithane). Follow instructions on the fungicide label.

DIPLODIA TIP BLIGHT OF AUSTRIAN, RED AND SCOTCH PINE

This disease is most commonly seen on Austrian pine and some of the other two-and three-needle pines such as red pine, Mugho pine and Scots pine. It is found more uncommonly on white pine, spruces and other evergreens. The fungus commonly attacks mature trees that have been under stress from drought, root restriction or other planting site problems. It can also be a problem in young, rapidly growing nursery or Christmas tree plantings.

Symptoms

The pathogen infects and kills current year's shoots. When the infected, current season's needles are 1/2 to 3/4 expanded, they turn yellow, then brown as they die on individual branch tips. A close look at the bases of the dead needles may reveal tiny black, fungal fruiting bodies emerging from the needle surface. Repeated infection over several years causes the ends of affected branches to have a proliferation of shoots. If left unchecked the disease can eventually kill mature trees. Girdling cankers can be formed if the pathogen infects wounds on the stem. Other problems can cause similar dieback and tree decline. Winter drying; drought; injury from weevils, pine-shoot moths or tip moths; and some needlecast diseases caused by other fungi may cause damage similar to tip blight.

Causal Fungus and Disease Development

Tip blight is caused by the fungus, Sphaeropsis sapinea,

once known as Diplodia pinea. Spores of the fungus develop in the black fruiting bodies located at the base of infected needles and other affected plant parts from spring through fall. They are spread about only during periods of rainfall. Pine shoots are particularly susceptible to infection in early spring.

Developing cone scales are also commonly infected, although they are not damaged. Wounds, such as those made by hail, shearing, or insects (weevil or spittlebug feeding) also serve as entry points for the fungus. The fungus survives over winter in the infected shoots, bark, cones or needle litter beneath the tree.

Control

- Trees should be kept in good vigour with regular maintenance, deep watering during droughts, fertilizing, control of insects and vertical mulching to open up the soil in the root zone. Vertical mulching can be done to improve landscape soils. Vertical mulching will lessen damage due to excessive water, preserve necessary aeration during wet periods, allow sub-soil water penetration during dry periods, and promote the formation of fine feeder roots.
  Drill one or two inch wide, 18" deep holes in the soil on 12-20" centres under affected trees near the drip line of the branches (where fine feeder roots are located). Fill holes with a mixture of equal parts of peat and a coarse aggregate such as pumice or calcined clay particles.
- Remove previously blighted shoots. Since many spores are produced on cones, removal of previously blighted shoots probably does not decrease spore numbers appreciably. However, it does serve to make the tree look better and may increase its vigour.
- Do not shear or prune infected trees during wet weather because spores released at this time may be carried from tree to tree on pruning tools.
- This disease can be partially controlled with fungicides. Attention must be given to protecting the

new spring growth of the trees from bud swell to full candle elongation.

Make first application just prior to bud break and make two or more additional applications at 10-day intervals. It is important to get the first application on the trees before any bud sheaths have broken. If bud sheaths have broken, spraying with fungicides is a waste of time and money.

MAINTAINING HEALTHY RHODODENDRONS AND AZALEAS IN THE LANDSCAPE

Rhododendrons and azaleas, which are closely related, are among the most popular flowering shrubs. In many urban landscapes, these shrubs thrive and have relatively few serious health problems once they are established. However, they do have some "special requirements" that must be met to insure good health.

These special conditions match those in areas where rhododendrons and azaleas are native. Rhododendrons and azaleas grow on forest floors in many parts of the world, in shaded habitats, with acidic soils rich in organic matter. Soils are often covered with a surface layer of decaying leaf litter.

Matching these conditions where native rhododendrons and azaleas thrive is the key to their good health in the landscape. Plant in areas with good soil drainage, acid pH and partial shade that are sheltered from direct afternoon sun and winter winds.

Some of the more common problems that can occur on rhododendrons and azaleas include: iron deficiency, winter injury (burn), black vine weevil, and *Phytophthora* root rot.

Iron Deficiency

The yellowing of rhododendron leaves typical of iron deficiency. The yellowing is between the veins and more severe on younger leaves. This problem generally results from plants growing in soils of improper pH.

Rhododendrons must be grown in acidic soil that is high in organic matter. If the pH is above 6.0, soil amendments such as sulfur, iron sulfate or ammonium sulfate must be incorporated into the root area to lower pH. It will be difficult to overcome the deficiency problem in soils high in lime or calcium, even with soil amendments. In such cases, mulch the plant heavily with a good grade of sphagnum peat. Bark mulch mixed with the peat provides a mulch with good aeration and drainage.

It also suppresses root rotting organisms. If kept moist, plants will root into this mulch. Use a complete, acid fertilizer that contains iron.

Winter Burn

Leaf drying and browning can occur on rhododendron leaves as a result of winter exposure. The leaves, even though they may be "rolled up" at times, are subject to drying out in dry winter air.

The solution is to protect the plant from the drying wind. Plant rhododendrons behind buildings or other plants that can serve as wind shields. Put wind shields in place around the plants during the winter months. Mulching as described above is critical to preventing winter injury.

Black Vine Weevil

The adult weevils feed on rhododendron leaves producing a C-shaped notching in the leaf margin. These insects can be quite damaging. Most of the damage comes from weevil larvae feeding on the roots. Affected plants lose vigour, and may die eventually.

The insecticide Orthene is registered for use on rhododendrons, and it will control black vine weevil adults.

Phytophthora Root Rot (Rhododendron Wilt)

This disease is caused by a soil-borne fungus (*Phytophthora* spp.). Generally, it is a problem where wet (saturated) soil conditions occur frequently. Early symptoms of the disease consist of retarded growth, drooping of foliage (perhaps on one or two branches only) and yellowing of leaves.

Infected roots appear dark and "mushy." As the disease

progresses, a browning discolouration of the wood may extend upward from the base on affected branches. Plants in poorly drained soils are very subject to waterlogging which makes them highly susceptible to this disease. If this occurs, plants may die quickly.

Whereas infected plants cannot be cured, root rot may be tolerated by the plant if improvements in soil drainage and aeration are made as soon as possible. Young plants can be lifted and replanted. Before replanting, improve the drainage and aeration of the soil.

Use tile drainage or add porous materials in a layer beneath the root zone. Plant in a raised bed and do not mound the soil up around the crown. Mulching with tree bark provides biological control. The mulch must be applied to a depth of two inches and reapplied as it decomposes.

If plants cannot be lifted and replanted, try to improve drainage and lessen the occurrence of over watering by redirecting rain runoff, placement of drain tiles, and changes in irrigation programmes.

If the plants die from root rot, it would be unwise to replant another rhododendron in the site without considerable improvement in the soil conditions. Caroline and English Roseum are rhododendron cultivars with some resistance to this disease.

Chapter 7

Vegetable Diseases

DAMPING-OFF AND ROOT ROT OF BEANS

Root rots, damping-off before and after seedling emergence, and seed rots are destructive diseases of green, snap, lima, and dry beans. These diseases are caused primarily by soilborne fungi.

Significant losses may occur to susceptible varieties, especially if cool, wet weather conditions prevail for the first few weeks after seeding and then are followed by hot, dry weather. Disease incidence and severity often vary greatly, even in areas with a history of root rot. In the same growing season, it is not uncommon to lose a crop completely and then re-seed and experience no problems.

This situation results from changes in biological, environmental, and soil conditions. Since there are no commercially acceptable resistant varieties, growers should learn how to recognize these diseases and use a combination of management practices to minimize potential losses.

Symptoms

Damping-off before emergence results from fungal attack of germinating seeds and/or young seedlings while they are still in the ground. Infected seeds may fail to germinate, become soft and mushy, and finally disintegrate. Slightly darkened water-soaked lesions may be visible on stems of young seedlings. Infected areas enlarge quite rapidly, and seedlings may die shortly after infection, prior to emergence from soil.

Roots or stems of seedlings that have already emerged also can be attacked at or below the soil line resulting in damping-off.

Infected roots are usually discoloured or rotted and sometimes reddish-brown lesions develop on the tap root. Infected stem tissues are soft and colourless to dark-brown. Basal portions of invaded stems may be much thinner than the areas above the lesion, a condition called "wire stem." As a result, the seedling may fall over and die.

Damping-off is a major cause of poor stand establishment in bean plantings. Older plants can also be attacked by these fungi. Later infections are usually confined to roots, which may result in stunting, wilting, or plant death. To diagnose bean root rots, suspected plants should be carefully dug and washed, because pulling plants may leave tissues with characteristic symptoms in the soil. If plants are brought to a diagnostic clinic, they should be dug and left intact in soil.

Causal Organisms

Root rots, damping-off, and seed rots are caused primarily by soilborne fungi. Fusarium root rot is caused by Fusarium solani f. sp. phaseoli. This fungus is capable of surviving long periods in soil, even in the absence of beans, by the production of thick-walled resting spores. Its host is mainly green beans, but lima beans and garden peas are also susceptible.

Coarse-textured, acidic and poorly fertilized soils favour development of Fusarium root rot. A number of fungi in the genus Pythium are capable of inciting seed rot and seedling damping-off. This may vary with environmental conditions and developmental stage of the host.

These fungi can survive in soil for many years, either by producing thick-walled resting spores, or in a vegetative condition within crop residues left from previous years. They may also survive by attacking several other garden vegetable crops such as beets, cabbage, peas, melons, squash, and cucumbers.

Development of diseases caused by Pythium is more prominent under 1) high or low temperatures that are unfavourable to crop growth, 2) excess water and/or nitrogen, or 3) continued cropping of susceptible plants. Rhizoctonia solani is another soil-inhabiting fungus causing root rots and damping-off. This fungus overwinters free in the soil or within decayed plant tissues. Aside from bean, its host range includes beets, cabbage, lettuce, peas, pepper, tomatoes, and many others. Warm, relatively dry soil conditions favour disease development.

- Do not grow beans or other susceptible crops continually in the same location. Continuous cropping of susceptible plants will eventually lead to a buildup of these fungi in the soil. Since they are capable of long-term survival, a rotation of 4-5 years is desirable. Avoid planting beans in fields known to be heavily infested with bean root-rot fungi.
- Plant beans only on well-drained soils or try to improve drainage. This could be done by improving soil structure and/or installing drain tiles. Subsoiling to a depth below the plowed layer will reduce soil compaction, and improve drainage. Deep plowing of the previous years' crop residues will reduce bean root rot.
- Delay planting until the soil is warm (above 65 F) and seed shallow to insure rapid emergence. Avoid planting seeds too close together-follow instructions on the seed container. Do not overfertilize, especially with nitrogen.
- Use of fungicide-treated seeds will minimize problems with damping off and root rots.

POWDERY MILDEW OF VINE CROPS

Powdery mildew can be a serious problem on foliage and stems of cucumbers, melons, squash, pumpkins, and ornamental gourds.

Symptoms

Small patches of fine, white threads develop on surfaces of infected leaf blades in late July on most vine crops. These patches grow together and eventually cover stems and foliage with white, powdery masses of spores. In severe infections, leaves will yellow and die.

Petioles, stems and, rarely, fruit will also become infected. Early death of leaves can decrease the total amount of fruits formed as well as reduce fruit size. Loss of foliage hastens maturity of fruit and increases sunburning. Stems infected with Powdery Mildew become brown. This can spoil the appearance of the "handles" on pumpkins which lowers value of the fruit.

Causal Organisms

Two fungi cause powdery mildew on vine crops, Sphaerotheca fuliginea and Erysiphe cichoracearum. S. fuliginea is most common. Powdery mildew fungi thrive under conditions of high relative humidity, warm temperatures, low light, high fertility, and succulent plant growth. Unlike bacteria and many fungi, free moisture on leaf surfaces actually inhibits infection by these pathogens, although very high relative humidity is required for spore germination.

Powdery mildew fungi grow only on living host plants. They survive the winter as dormant mycelium on perennial plants or as spores in thick-walled fruiting structures. Powdery Mildew spores can be blown into the spring from warmer southern areas. Mature foliage is most readily infected; very young leaves are nearly immune.

Control

- Plant resistant cultivars of cucumbers and muskmelons. Pumpkin varieties vary greatly in susceptibility to this disease.
- Use cultural practices that avoid excessive succulence, overcrowding, shading, overwatering, or excess fertilization especially with nitrogen.
- Avoid making new planting of vine crops in the vicinity of older plantings, especially if mildew is already present.
- Beginning in late July or early August when Powdery

Mildew first appears, regular applications of fungicides may be required.

Common disease of both potato and tomato is Early blight is a very. It causes leaf spots and tuber blight on potato, and leaf spots, fruit rot and stem lesions on tomato. The disease can occur over a wide range of climatic conditions and can be very destructive if left uncontrolled, often resulting in complete defoliation of plants. In contrast to the name, it rarely develops early, but usually appears on mature foliage.

Symptoms

On leaves of both crops, the first symptoms usually appear on older leaves and consist of small, irregular, dark brown to black, dead spots ranging in size from a pinpoint to 1/2 inch in diameter. As the spots enlarge, concentric rings may form as a result of irregular growth patterns by the organism in the leaf tissue. This gives the lesion a characteristic "target-spot" or "bull's eye" appearance.

There is often a narrow, yellow halo around each spot and lesions are usually bordered by veins. When spots are numerous, they may grow together, causing infected leaves to turn yellow and die. Usually the oldest leaves become infected first and they dry up and drop from the plant as the disease progresses up the main stem.

On tomato, stem infections can occur at any age resulting in small, dark, slightly sunken areas that enlarge to form circular or elongated spots with lighter-coloured centres. Concentric markings, similar to those on leaves, often develop on stem lesions. If infested seed are used to start tomato transplants, seedlings may damp off soon after emergence.

When large lesions develop at the ground line on stems of transplants or seedlings, the plants may become girdled, a condition known as "collar rot." Such plants may die when set in the field or, if stems are weakened, may break over early in the season.

Some plants may survive with reduced root systems if portions of stems above the canker develop roots where they contact the soil.

Such plants, however, usually produce few or no fruits. Stem lesions are much less common and destructive on potato.

Blossom drop and spotting of fruit stems, along with loss of young fruit, may occur when early blight attacks tomatoes in the flowering stage. On older fruits, early blight causes dark, leathery sunken spots, usually at the point of stem attachment.

These spots may enlarge to involve the entire upper portion of the fruit, often showing concentric markings like those on leaves. Affected areas may be covered with velvety black masses of spores. Fruits can also be infected in the green or ripe stage through growth cracks and other wounds. Infected fruits often drop before they reach maturity.

On potato tubers, early blight results in surface lesions that appear a little darker than adjacent healthy skin. Lesions are usually slightly sunken, circular or irregular, and vary in size up to 3/4 inch in diameter. There is usually a well defined and sometimes slightly raised margin between healthy and diseased tissue. Internally, the tissue shows a brown to black corky, dry rot, usually not more than 1/4 to 3/8 inch deep.

Causal Organism

Early blight is caused by the fungus, Alternaria solani, which survives in infected leaf or stem tissues on or in the soil. This fungus is universally present in fields where these crops have been grown. It can also be carried on tomato seed and in potato tubers. Spores form on infested plant debris at the soil surface or on active lesions over a fairly wide temperature range, especially under alternating wet and dry conditions. They are easily carried by air currents, windblown soil, splashing rain, and irrigation water.

Infection of susceptible leaf or stem tissues occurs in warm, humid weather with heavy dews or rain. Early blight can develop quite rapidly in mid to late season and is more severe when plants are stressed by poor nutrition, drought, or other pests. Infection of potato tubers occurs through natural openings on the skin or through injuries. Tubers may come in contact with spores during harvest and lesions may continue to develop in storage.

- Use a crop rotation that includes potatoes or tomatoes only every third or fourth year to allow infested plant debris to decompose in the soil. Rotations with small grains, corn or legumes are preferable.
- Use tillage practices such as fall plowing that bury all plant refuse.
- Select cultivars that have a lower susceptibility to early blight.
- Use certified disease-free tomato seed and transplants.
- If producing tomato transplants, disinfest soil in plant beds' and control humidity in cold frames or greenhouses. Practice good sanitation throughout the transplant production operation.
- Use appropriate measures to control weeds and volunteer potatoes and tomatoes in production areas.
- Maintain fertility at optimal levels-nitrogen and phosphorus deficiency can increase susceptibility to early blight.
- Time applications of overhead irrigation to allow plants to dry before nightfall.
- Although the above measures are important to minimize infection, it is usually necessary to apply fungicide sprays to fully protect plants from early blight. Applications to tomato are usually begun 2-3 weeks following emergence or soon after transplanting if a calendar schedule is followed. Timing of fungicides can also be made using the TOM-CAST system. For potatoes, fungicide applications should be initiated when plants begin to flower. Thorough coverage is important. Adjust equipment for good vine penetration.

Late Blight of Potato and Tomato

Late blight is one of the most devastating diseases of potato and tomato worldwide. It was responsible for the devastating Irish potato famine of the 1840's and has continued to be important to the present. Since 1990, late blight has caused widespread damage across the United States and Canada. If left unmanaged, this disease can result in complete destruction of potato or tomato crops.

Symptoms

Late blight appears on potato or tomato leaves as pale green, water-soaked spots, often beginning at leaf tips or edges. The circular or irregular leaf lesions are often surrounded by a pale yellowish-green border that merges with healthy tissue. Lesions enlarge rapidly and turn dark brown to purplish-black.

During periods of high humidity and leaf wetness, a cottony, white mold growth is usually visible on lower leaf surfaces at the edges of lesions. In dry weather, infected leaf tissues quickly dry up and the white mold growth disappears. Infected areas on stems appear brown to black and entire vines may be killed in a short time when moist weather persists.

On potato tubers, late blight appears as a shallow, coppery-brown, dry rot that spreads irregularly from the surface through the outer 1/8-1/2 inch or more of tissue. On tuber surfaces, lesions appear brown, dry, and sunken, while infected tissues immediately beneath the skin appear granular and tan to copper-brown.

When tubers are stored under cool, dry conditions, lesion development is retarded and, upon prolonged storage, lesions may become slightly sunken and desiccated. Secondary bacteria and fungi frequently enter late-blight lesions, usually resulting in a slimy breakdown of entire tubers.

Late blight can also develop on green tomato fruit, resulting in large, firm, brown, leathery-appearing lesions, often concentrated on the sides or upper fruit surfaces. If conditions remain moist, abundant white mold growth will develop on the lesions and secondary soft-rot bacteria may follow, resulting in a slimy, wet rot of the entire fruit.

Causal Organism

Late blight is caused by the fungus *Phytophthora infestans*. Unlike most pathogenic fungi, the late blight fungus cannot survive in soil or dead plant debris. For an epidemic to begin in any one area, the fungus must survive the winter in potato tubers (culls, volunteers), be reintroduced on seed potatoes or tomato transplants, or live spores must blow in with rainstorms.

Disease development is favoured by cool, moist weather. Nights in the 50's and days in the 70's accompanied by rain, fog or heavy dew are ideal. Under these conditions, lesions may appear on leaves within 3-5 days of infection, followed by the white mold growth soon thereafter.

Spores formed on the mold are spread readily by irrigation, rain and equipment. They are easily dislodged by wind and rain and can be blown into neighbouring fields within 5-10 miles or more, thus beginning another cycle of disease.

Infection of potato tubers arises from spores that develop on foliage. Tubers exposed by soil cracking or erosion of hills may come in contact with spores washed down from infected leaves and stems by rainfall or irrigation. Tubers infected during the growing season may partially decay before harvest. Tuber infection may also occur at harvest when tubers contact living spores remaining on infected vines. Little if any tuber-to-tuber spread of late blight occurs during storage if tubers are kept under cool, well-ventilated conditions.

The protectant fungicides commonly used to protect plants from late blight remain fully effective with all known strains of the fungus.

- Infected cull potatoes are a major source of spores of the late blight fungus and must be disposed of properly-Do Not Make Cull Piles. Cull potatoes should be spread on fields not intended for potato production the following year in time that they will totally freeze and be destroyed during the winter. If this is not possible, they must be destroyed in some other way such as by complete chopping, burial, burning or feeding to livestock.
- Plant only certified seed potatoes. Use of "year-out" seed or seed saved from local crops is asking for trouble with late blight. Seed sources should be

selected very carefully to avoid bringing in late blight on seed potatoes, especially new strains of the fungus. Look for the characteristic coppery-brown discolouration of the potato flesh under the skin of seed tubers.

Infected tomato transplants also can be a significant source of the disease. Use only obviously healthy tomato transplants free of dark lesions on leaves or stems.

- Volunteer potatoes and tomatoes can be a significant source of spores of the late blight fungus. All volunteers should be destroyed as quickly as possible by herbicides, chopping, or cultivation.
- Growers should scout fields regularly to look for late blight. Special attention should be paid to early-planted fields because that is where the disease is likely to develop first.Scouting should be concentrated in low-lying areas, field edges along creeks or ponds, near the centre of centre-pivot irrigation rigs, in areas near woodlots or any area that is protected from wind where the leaves tend to remain wet longer.

  Any area where it is difficult to apply fungicides such as edges and corners or under power lines if using aerial application should be examined. Scouts should look for large, black or purplish lesions on stems or leaves and the telltale cottony, white mold growth, usually on the undersides. Be sure to check leaves and stems under the crop canopy as that is where the disease is most likely to begin.
- Use of a good protectant fungicide programme is necessary to fully protect any crop of potatoes or tomatoes.
- With potatoes, make sure that vines have been completely dead for 2-3 weeks prior to harvest. Fungicide applications should be continued until vines are dead. When foliage dies, spores of the late blight fungus that remain on the foliage also die.

This practice will prevent infection of tubers during harvest and development of late blight in storage.

BACTERIAL RING ROT OF POTATOES

Bacterial ring rot is an important disease of potatoes and is one of the main reasons for rejection of seed potatoes from certification programmes.

This disease is particularly serious because it has the potential to spread quickly throughout a farm and may lead to severe losses if left unchecked. Ring rot was originally found in Germany in the late 1800's. The causal bacteria were introduced into the United States in the early 1930's and by 1940 were found throughout the country.

Symptoms

Severe ring rot can result in wilting of leaves and stems along with yellowing and death of leaves. Lower leaves usually wilt first, are slightly rolled at the margins, and are paler green than healthy leaves. As wilting progresses, leaf tissues between veins become yellow. In the later stages of disease, margins of lower leaves die and become brittle, and eventually entire stems yellow and die.

Frequently, only one or two stems in a hill will develop symptoms and, in some cases, there are no above-ground symptoms at all. Ring rot derives its name from a characteristic breakdown of the vascular ring within the tuber. This often appears as a creamy-yellow to light-brown, cheesy rot. The symptom is most frequently observed when a diseased tuber is cut crosswise at the stem end.

In severe cases, the vascular ring may be separated, and a creamy or cheesy exudate can be forced out from this tissue when the tuber is squeezed. On the outer surface, severely diseased tubers may show slightly sunken, dry, cracked areas. Infected tubers are often invaded by secondary decay organisms which may lead to complete breakdown.

Symptoms of ring rot in the vascular tissue of infected tubers are often less obvious than described above, appearing as only a broken, sporadically appearing dark line, or as a continuous, yellowish discolouration. Because of this, laboratory tests should always be performed to confirm a diagnosis of ring rot.

Causal Organism

Ring rot is caused by the bacterium Clavibacter michiganense subsp. sepedonicus. Ring-rot bacteria survive between seasons mainly in infected seed tubers. They are also capable of surviving 2-5 years in dried slime on surfaces of crates, bins, burlap sacks, or harvesting and grading machinery, even if exposed to temperatures well below freezing.

Survival is longest under cool, dry conditions. Ring-rot bacteria do not survive in soil in the absence of potato debris, but can survive from season to season in volunteer potato plants. Wounds are necessary for penetration of the bacteria into seed pieces.

The pathogen is easily transmitted from diseased tubers to healthy seed pieces during the seed-cutting process. A knife that cuts one infected tuber can spread these bacteria to the next 20-100 seed pieces. Likewise, the bacteria may be spread during planting, particularly if a picker-type planter is used.

Ring-rot bacteria can be moved in irrigation water and by chewing insects, such as Colorado potato beetles and flea beetles. After the bacteria become established in a plant, they multiply and move throughout the plant via the water-conducting tissues. Fortunately ring-rot bacteria are capable of causing disease only in potato, although they may be able to colonize roots of sugar beets.

- Plant only certified disease-free seed tubers. Certified seed potatoes are produced under regulations mandating zero tolerance for ring rot. Although use of certified seed tubers will not guarantee total freedom from ring rot bacteria, it is the best assurance.
- Discontinue use of any lot of seed tubers in which ring rot is found. Seed lots known to be contaminated with ring-rot bacteria should never be planted.

- Before handling seed tubers, all containers, tools, knives and mechanical cutters, planters, and other equipment should be thoroughly washed with a detergent solution, rinsed, and then sanitized with a disinfectant (for current recommendations,
- If ring rot is confirmed to be present, a thorough cleanup must be undertaken. Dispose of all infected tubers away from potato production areas. Clean all surfaces of storages and equipment to remove all mud, dirt and debris and then wash with a strong detergent in hot water applied by a high-pressure washer.

  After cleaning, sanitize all storages and equipment with a disinfectant. Do not plant potatoes for two seasons in any field in which ring rot has been found.

POTATO PINK ROT, PYTHIUM LEAK AND SEED-PIECE DECAY

Pink rot and Pythium leak, sometimes collectively called water rot, occur sporadically wherever potatoes are grown. These diseases are a problem of mature tubers at harvest and in storage. They are most serious when warm, wet soil conditions persist during tuber formation and at harvest.

When newly-planted seed pieces are exposed to these conditions, Pythium seed-piece decay also can be severe. Major problems with these diseases are usually associated with excessive rainfall or irrigation either early or late in the season, especially on poorly-drained soils.

Symptoms

Pythium seed-piece decay often results in delayed emergence and poor stands. Infected seed pieces become a soft, watery mass in the soil. Symptoms of pink rot in mature plants include brown or blackened roots or stolons, and in severe cases, leaf chlorosis, stunting, wilting, and even plant death. Tubers develop pink rot mostly through diseased stolons, but occasionally infections occur at buds or lenticels.

Decay spreads through infected tubers with the advancing

margin of the rot usually sharply defined by a dark line, which may be visible through the skin. Eyes of infected tubers are often dark brown. Decaying tubers remain intact, but are spongy and odorless. If squeezed, a clear liquid will exude. When infected tubers are cut open, the internal tissues turn salmon pink after a 15-20 minute exposure to air, then later become brownish-black.

Pythium leak usually develops through harvest wounds in tuber surfaces and begins as a discoloured, watersoaked area. As with pink rot, the advancing margin of infection is usually bounded by a dark line. Infected tissues are extremely watery, and appear brown or gray. Severely rotted tubers are of a uniform texture resembling a soft, watery paste.

Causal Organisms

Pink rot is caused by several species of the soilborne fungus Phytophthora (NOT the species that causes late blight) while Pythium leak and seed-piece decay are caused by several species of the closely related soilborne fungus Pythium. These fungi are widely distributed in both water and soil, and their behaviour is similar.

They survive in soil within decaying plant material, or as resistant spores free in the soil. In warm, moist soil, these fungi produce swimming spores that move in water films. Roots can be infected by Phytophthora at almost any stage of plant growth, but symptoms are more severe on younger roots. Both groups of fungi infect tubers through wounds, but Phytophthora generally infects tubers before harvest, often through stolons.

Infection by Pythium usually occurs through harvest wounds, especially at temperatures above 70 F. Seed pieces can be infected by Pythium as soon as they are planted.

- Select areas with well-drained soils for planting potatoes.
- Use a crop rotation away from potatoes for at least 4 years if pink rot or leak have been severe. This may reduce the amount of fungus surviving in the soil.
- Delay planting for at least 2 weeks after plowing

down green vegetation as this may temporarily stimulate populations of Pythium fungi.

- Avoid planting in soils colder than 45 F or warmer than 70 F.
- Avoid harvesting infested fields when soils are especially wet or soil temperatures are below 50 F or above 65 F. Stop irrigation well in advance of harvest.
- Avoid bruising tubers during harvest by adjusting equipment properly, keeping digger chains fully loaded and minimizing drops to 6 inches or less. Do not leave harvested tubers lying on warm, moist soils for any length of time as infection with Pythium may occur quickly.
- Leave low spots in fields unharvested if they have been waterlogged and much rot is present.
- Keep tubers cool and as dry as possible during harvest, loading, transit and storage.
- Grade out infected tubers as much as possible before placing harvested tubers in storage.
- Store lots of harvested tubers containing many infected tubers separately from healthy lots. Good airflow through the pile should be provided to dry out leaky tubers. Lots with significant amounts of disease should be marketed as soon as possible as they will not store well.

SCAB OF POTATO TUBERS

Scab is a disease of potato tubers that results in lowered tuber quality due to scab-like surface lesions. There are no above-ground symptoms. Two forms of scab occur. Common scab occurs in all production areas and is most severe in soils with a pH above 5.5. Another less common form, called acid scab, is important in acidic soils (below pH 5.5).

Symptoms

Scab symptoms are quite variable. Usually, roughly circular, raised, tan to brown, corky lesions of varying size develop randomly across tuber surfaces. Sometimes scab develops as a rather superficial layer of corky tissues covering large areas of the tuber surface. This is called russet scab. Pitted scab can also occur where lesions develop up to 1/2 inch deep.

These deep lesions are dark brown to black, and the tissues underneath are often straw-coloured and somewhat translucent. More than one of these lesion types may be present on a single tuber.

Although scab symptoms are usually noticed late in the growing season or at harvest, tubers are susceptible to infection as soon as they are formed. Small brown, water-soaked, circular lesions are visible on tubers within a few weeks after infection.

Mature tubers with a well-developed skin are no longer susceptible, but existing lesions will continue to expand as tubers enlarge. Thus disease severity increases throughout the growing season. Scab is most severe when tubers develop under warm, dry soil conditions. Coarse-textured soils that dry out quickly are therefore more conducive to scab than are fine-textured soils.

A few other conditions can be confused with scab. White, enlarged lenticles, which frequently occur on potato tubers harvested from wet soil, can be mistaken for scab. Usually this condition will disappear when tubers are dried. Patchy russeting, checking, or cracking of tuber surfaces caused by the fungus Rhizoctonia also may be confused with russet scab. A totally different but uncommon disease called powdery scab, caused by the fungus Spongospora subterranea, causes very similar scab-like symptoms. Laboratory examination may be necessary to identify these diseases.

Causal Organisms

Scab is caused by a group of filamentous bacteria called actinomycetes that occur commonly in soil. In soils with a pH above 5.5, Streptomyces scabies is usually responsible for common scab, and is capable of causing all the types of scab lesions described above. It is commonly introduced into fields on seed potatoes, and will survive indefinitely on decaying plant debris once the soil is contaminated. Because the organism can survive passage through the digestive tract of animals and be distributed.

Blackleg, Aerial Stem Rot, and Tuber Soft Rot of Potato

Blackleg, aerial stem rot, and tuber soft rot are all similar diseases caused by several types of soft-rot bacteria. Blackleg and tuber soft rot occur wherever potatoes are grown. Aerial stem rot is also widespread, but is most severe under sprinkler-irrigation.

Symptoms

Blackleg begins from a contaminated seed piece, but the symptoms can occur at several stages of plant development. In severe cases, entire seed pieces and developing sprouts may rot in the ground prior to emergence, resulting in a poor stand. Blackleg often develops after plants are well up or even in flower.

In this case, stem bases of diseased plants typically show an inky-black to light-brown decay that originates from the seed piece and can extend up the stem from less than an inch to more than two feet. Leaves of infected plants tend to roll upward at the margins, become yellow, wilt, and often die.

Aerial stem rot (also called bacterial stem rot or aerial blackleg) is initiated by soft-rot bacteria from sources external to the seed piece. Stem infection can occur through wounds or through natural openings such as leaf scars. Lesions on diseased stems first appear as irregular brownish to inky-black areas.

These enlarge into a soft, mushy rot that causes entire stems to wilt and die. Potato tubers with soft rot have tissues that are very soft and watery, and have a slightly granular consistency. The diseased tissue is cream- to tan-coloured, and often has a black border separating diseased from healthy areas.

In the early stages, soft-rot decay is generally odorless, but later a foul odor and a stringy or slimy decay usually develops as secondary decay bacteria invade infected tissues.

Most internal tuber tissues may be consumed by soft rot organisms, sometimes leaving only a shell of skin remaining in the soil.

Causal Organisms

Blackleg, aerial stem rot, and tuber soft rot are caused by two closely related bacteria, Erwinia carotovora subsp. atroseptica and Erwinia carotovora subsp. carotovora. E. c. carotovora is very common and has an extensive host range, including most fleshy vegetables.

It survives readily in soil and surface waters such as rivers, lakes, and even oceans. These bacteria are capable of multiplying and persisting in the root zones of many host and nonhost crop and weed species. In contrast, E. c. atroseptica is associated mostly with potatoes.

These bacteria do not survive well in soil for more than one year, unless they are contained within diseased tubers or other potato plant debris. Blackleg is usually caused by E. c. atroseptica carried on contaminated seed tubers.

Most lots of seed tubers are contaminated to some degree, but the bacteria are usually dormant and do not cause disease unless environmental conditions are favourable. In contrast, aerial stem rot is usually caused by E. c. carotovora contained in infested soil or introduced to the crop by irrigation water, wind-blown rain, and insects. Tuber soft rot can be caused by either of these soft-rot bacteria.

Moisture and temperature are the two critical factors in initiation and development of soft-rot diseases. High soil temperatures and bruising of seed tubers favour seed-piece decay and pre-emergence blackleg. Blackleg in growing plants is favoured by cool, wet soils at planting followed by high temperatures after emergence.

Dense plant canopies and long periods of leaf wetness favour infection of aerial plant parts. Although tuber soft rot can occur at any temperature above 50 F, disease develops best above 75 F. Oxygen depletion in tubers also favours soft rot. When seed pieces in soil or tubers in storage become covered with a film of water, the tissues rapidly become depleted of oxygen. This also may be induced by soil flooding or improper drying of washed tubers.

Once it starts, tuber soft rot can proceed rapidly in storage. "Wet" areas may develop in the piled tubers that flow onto ones below, spreading the bacteria. Heat, coupled with condensation on tuber surfaces, can further adversely affect storage conditions, resulting in accelerated "melt" of the pile.

- Plant only certified, disease-free seed tubers. If possible, use whole (B-size) seed tubers that do not have to be cut.
- When receiving seed tubers in bags, do not stack more than five bags high. With bulk or bagged seed, store at 40-45 F until 2-3 weeks before planting, then warm to 55-60 prior to cutting.
- Clean all equipment used for cutting seed tubers thoroughly and then sanitize with an appropriate disinfectant.
- Treat cut seed pieces with recommended fungicide dressings immediately after cutting.
- Plant treated cut seed pieces immediately if soil temperatures are 55-65 F at planting depth. Seed pieces can be held 1-2 weeks at 55-60 F and 95-99 per cent relative humidity to hasten healing of cut surfaces. Condensation on surfaces of seed pieces must be avoided.
- Do not irrigate fields until plants are well emerged. Avoid using surface water for irrigation.
- During crop growth, monitor irrigation and nitrogen fertility to minimize excessive vine growth that will promote leaf wetness within the plant canopy.
- Harvest tubers only after the vines are completely dead to ensure skin maturity. Low spots in the field should be left unharvested if significant waterlogging has occurred.
- Take all precautions to minimize cuts and bruises when harvesting and handling tubers.
- Hold newly harvested potatoes at 55-60 F with 90-95 per cent relative humidity for the first 1-2 weeks to

promote wound healing. After this curing period, lower the temperature of table stock to 38-40 F for long-term storage. Never wash tubers prior to storage.

FUSARIUM DRY ROT AND SEED-PIECE DECAY OF POTATO

Fusarium dry rot is an important postharvest disease of potato tubers that causes significant losses in storage and transit of both seed tubers and those for table consumption. It is also a major cause of seed-piece decay after planting.

Symptoms

Infected tubers usually develop a dry rot, but a moist rot may occur if secondary infections with soft-rot bacteria also are involved. Surfaces of infected tubers are sunken or wrinkled, and rotted tissues appear brown or gray to black. A white or pink mold is sometimes visible on tuber surfaces. When tubers are cut, internal cavities within rotted tissues may contain.

White, yellow or pink molds. In storage, blue, black, purple, gray, white, yellow, or pink spore masses may develop in these internal cavities. After low-temperature storage, internal tissues often will become firm and dry or even powdery.

Causal Organisms

Fusarium dry rot is caused by several species of the soilborne fungus Fusarium. These fungi are common in most soils where potatoes are grown and survive as resistant spores free in the soil or within decayed plant tissues. Although some infections may develop on tubers before harvest, most infections occur as the fungus enters tubers through harvest wounds. Small, brown lesions appear at wound sites 3-4 weeks after harvest and continue to enlarge during storage, taking several months to develop fully.

The disease develops fairly rapidly at temperatures above 50 F, but lesions will cease enlarging below 40 F. The fungus is only dormant at these low temperatures, however, and will resume growth when tubers are warmed. Fusarium seed-piece decay is really the same disease. Seed tubers may be infected prior to shipment. Decay during transit or storage often accounts for poor quality seed tubers.

When these Fusarium fungi are present on seed pieces or in the soil, poor stands may result, especially if cut surfaces of seed pieces are not properly healed. Fusarium seed-piece decay begins as reddish-brown to black depressions on cut surfaces. These may expand to cover the entire seed piece and often result in a slimy rot when infection by secondary soft-rot bacteria follows.

- Harvest tubers only after the vines are completely dead to ensure skin maturity.
- Take all precautions when harvesting and handling tubers to minimize cuts and bruises.
- Hold newly harvested potatoes at 55-60 F with 90-95 per cent relative humidity for the first 1-2 weeks to promote wound healing. After this curing period, lower the temperature of table stock to 38-40 F for long-term storage.
- Plant only certified, disease-free seed tubers. If possible, use whole (B-size) seed tubers that do not have to be cut into seed pieces before planting.
- When receiving seed tubers in bags, do not stack more than five bags high. With bulk or bagged seed, store at 40-45 F until 2-3 weeks before planting. Then allow seed potatoes to warm prior to cutting.
- Treat cut seed pieces with recommended fungicide dressings immediately after cutting.
- Plant treated cut seed pieces immediately or store them at 55-60 F and 95-99 per cent relative humidity to hasten healing of cut surfaces. Condensation on surfaces of seed pieces must be avoided.

RHIZOCTONIA STEM AND STOLON CANKER OF POTATO

Rhizoctonia stem and stolon canker, also called black

scurf, is an extremely common problem on potatoes that can result in delayed emergence, reduced stands and poor tuber quality.

Symptoms

Rhizoctonia canker often goes unnoticed until harvest when tubers are found to be covered with small brownish-black fungal bodies (sclerotia) that look somewhat like bits of black soil that will not wash off. These may vary in size from a pinhead to as large as a pea. When introduced into soil, these structures can survive and germinate the following spring to attack young shoots, roots, stolons, and tubers of the new crop. In many cases, weak, spindly-looking or late-emerging plants may be the result of an attack by this fungus.

The first sprouts are often killed before they reach the surface, resulting in the emergence of a weaker secondary sprout. Dry, sunken, brownish lesions developing on the base of the stem below the soil line are evidence of the stem canker phase of this disease. The lesions may girdle the stem or large cankers may interfere with movement of nutrients from the leaves to the tubers. In such cases, vines become yellow to reddish purple; the leaves begin to curl upward; the stalks swell, particularly at the nodes; and often small purplish tubers form where leaves branch from stems.

During midseason under a dense canopy of foliage, the fungus may develop a white, powdery mold growth on stems, extending just above the. soil line. Cankers that form on stolons may prune off young developing tubers. This disease may lead to russeting or surface cracking of mature tubers and sometimes shallow, brown lesions will form around lenticles. Low soil temperatures in the fall favour formation of sclerotia on tubers.

Causal Organism

This disease is caused by the fungus Rhizoctonia solani which can survive in the soil for many years, even under relatively dry conditions. Rhizoctonia causes disease in a wide variety of crops, but the strains found in association with potato generally do not attack and reproduce on other plant species. The fungus survives in soil associated with decomposing plant residues. In addition, the sclerotia can survive on infected tubers and can persist free in soil for extended periods. Emerging sprouts are usually attacked by fungus present on the seed tubers.

Once green leaves develop on sprouts, stem tissues are much less susceptible to infection. Stem cankers, stolon infections, and sclerotia on tubers usually develop when these tissues grow in proximity to sources of the Rhizoctonia fungus in soil. Cool (55-60 F), moist soils are optimal for infection. Sclerotia form on the surfaces of mature tubers under cool, moist conditions, generally after the vines have begun to die.

- Use a crop rotation with corn, grasses, and cereal grains. If this disease has been severe, 3-5 years should elapse between potato crops.
- Plant certified seed tubers that are free of Rhizoctonia on the skin.
- Use planting practices that promote rapid emergence: Avoid planting in heavy, poorly-drained soils. Plant seed potatoes when soil is warm (above 60 F). Cover seed tubers with no more than 2 inches of soil.
- Harvest tubers promptly after vines are dead to avoid the development of sclerotia on the surfaces of tubers while still in the soil.

MOSAIC VIRUS DISEASES OF VINE CROPS

Mosaic diseases of vine crops are caused by at least five different viruses: 1) cucumber mosaic virus (CMV), 2) squash mosaic virus (SQMV), 3) watermelon mosaic virus-2 (WMV-2), 4) zucchini yellow mosaic virus (ZYMV) and 5) papaya ringspot virus-watermelon isolate (PRSV-W). All five of these viruses can infect all cultivated vine crops (squash, melons, gourds, cucumbers, pumpkins) and under ideal conditions can cause a high rate of crop failure and severe economic losses.

Symptoms

As the names imply, all five viruses cause a mottling of

foliage called mosaic. This is characterized by the presence of intermingled patches of normal and light green or yellowish coloured plant tissue.

Depending on environmental conditions, mosaic symptoms can range from mild to severe and be visible on both leaves and fruit. The younger the plant when infected, the more severe the symptoms as the plant matures. In some cases, plants infected at the seedling stage may collapse and die. Plants infected at the flowering stage may not set fruit or young fruits may abort.

If plants are more mature when infected they do not show severe mosaic and may still produce marketable fruit. The most dramatic symptoms are often associated with infected fruit. Fruit symptoms can range from subtle colour change to severe deformation. It is not uncommon to have two or more viruses infecting the same plant and in these cases symptoms may be much more severe than if the plant were infected with only one virus.

It is nearly impossible to distinguish between any of the five viruses based only on visible symptoms. Because of this, it is important to have suspect plants sent to a diagnostic clinic capable of using specialized techniques to identify the viruses that are present. Proper disease management depends on knowing which specific virus is involved.

Disease Cycle and Transmission

All five vine crop mosaic viruses are transmitted by insects. CMV, ZYMV, WMV-2 and PRSV-W are transmitted primarily by the green peach aphid and the melon aphid. SQMV is transmitted by the striped cucumber beetle and spotted cucumber beetle and also through seed, but only at a very low rate. In most cases these insects acquire the virus by feeding on weeds or other vine crops that are infected. The insects then feed on noninfected vine crops and in the process of this feeding transmit the virus to the host plant. In the case of the green peach and melon aphid, the insect carries the virus on its probing mouthparts and the aphid only needs to probe the plant to transmit the virus.

Feeding by the aphid is not necessary. Once the insect probes one plant, it can move onto the next, probing and infecting as it goes. The more aphids present, the quicker and more complete the virus spread. By this process, large acreages can become infected quickly.

- Plant varieties with resistance to these viruses whenever possible. The availability of varieties with virus resistance varies depending on the virus and the vine crop in question. Major seed producers list resistant varieties in their seed catalogs.
- Control weeds which could harbor the viruses from areas around production fields. Pokeweed is a primary weed in the area which harbors many plant viruses. Weeds also serve as hosts for insects that transmit the viruses.
- Since the virus is moved about by insects it would seem logical that insect control would be a primary method of virus control. However, in this case it is not that simple. Since aphid vectors need only probe the plant to transmit the virus, insect control is not really that effective in controlling the virus.
  Even if the aphid were to die immediately following probing, virus transmission would still take place. To control the insect effectively insecticides would need to be applied on a daily basis. The economics of this may not be feasible. Aphids also move long distances with weather fronts making local control difficult.

SEPTORIA LEAF SPOT OF TOMATOES

Septoria leaf spot is a disease of the foliage and stems. It does not affect fruit directly. The disease causes rapid defoliation when weather is warm and moist.

Symptoms

Although the symptoms may appear on the leaves and stems at any stage of plant development, they usually become evident after plants have begun to set fruit in mid-July.

Initially the fungus causes small, water-soaked, roughly circular spots scattered randomly over the leaf. These spots enlarge to become 1/16 to 1/8 inch in diameter with dark margins and tan centres.

Small, dark, pimple-like fruiting bodies in which spores of the fungus are produced can be seen in the centre of the lesions. Older leaves near the ground usually show symptoms first. Symptoms appear rapidly on younger foliage in rainy weather. When leaves are heavily infected, they drop prematurely. Exposed fruits are subject to sunscald. The fungus which causes this disease can grow on weeds such as Jimson weed, horse nettle and nightshade.

Causal Organism

Septoria lycopersici lives between tomato crops in the soil on infested debris of tomato and weeds. Spores formed on crop debris splash onto foliage and start the disease. Wind and rain spread spores produced in the dark bodies formed in leaf spots to adjacent uninfected leaves. The fungus is most active between 60 and 80 degrees F when rainfall is abundant.

Control

- Rotate out of tomatoes for 4 years.
- Deep plow, preferably in the fall, to bury all plant refuse.
- Grow tomato transplants in sterilized soil.
- Control weeds, especially horse nettle, Jimson weed, and nightshade.
- Use of a protectant fungicide may be necessary to adequately control Septoria Leaf Spot when conditions are favourable for disease development.

FUSARIUM WILT OF VINE CROPS

Several different species of Fusarium, a soil-borne fungus, cause wilting of watermelon, muskmelon, cucumber, squash, and other vine crops. In many cases the fungus-causing wilt in a particular crop is specific to that crop. These fungi are generally capable of surviving for long periods in the soil

Wilting of older plants is often the first symptom of the disease. Before diseased plants totally collapse, however, they may begin to wilt during the hottest part of the day and recover during the night. Infected plants are often stunted and yellowed. Leaves often have dead areas which can mimic nutrient deficiencies. Stems of wilted plants when cut lengthwise at the soil line may show brown discolouration in the woody tissues immediately under the bark. Vines killed by Fusarium can be covered with pinkish-white fungal growth in wet weather.

Causal Organism

Fusarium oxysporum is well adapted to life in the soil. It can survive season to season in old diseased vines or it can live free in infested soil for many years in the absence of its host crop. The fungus grows at soil moisture and temperature favourable for vine crop growth.

If the soil is very wet, infection is reduced. The fungus is stimulated to germinate when roots of susceptible host plants are growing nearby. It enters the plant through root tips or where some opening is present.

Control

- Plant wilt-resistant cultivars whenever possible. Degree of resistance is influenced by the populations of the fungi in the soil, and which races are present.
- Avoid introduction of the fungus into fields. Once soil is infested, crop rotation will be of little use because of the long-term survival of these fungi in soil. The fungus can be spread on equipment, tools, feet, and surface water contaminated with infested soil. Do not put compost on fields which has been made from diseased vines. Compost resulting from such piles will contain the fungus.
- Crop rotations of three to four years may be helpful in lowering the amount of Fusarium in the soil. Since the types of Fusarium-causing wilt in a crop are generally specific to that crop it is possible to rotate between vine crops in infested soil.

- In commercial production of vine crops, soil fumigation may be useful.

ANTHRACNOSE OF TOMATO

Anthracnose is a common and serious disease of tomato fruit. This disease can occasionally cause severe damage to peppers, especially when red fruit is allowed to develop. Anthracnose can reduce a bountiful harvest into rotted fruit in a few days in warm, moist weather.

Symptoms

Small, watersoaked, circular lesions develop under the skin of fruit as it ripens. These become sunken and dark. Numerous dark specks, the fruiting bodies of the fungus, develop in the lesions in concentric rings. In moist, warm weather, these black bodies ooze gelatinous pink spore masses. In warm weather the fungus and soft rot bacteria which enter the split skin over the lesions spread internally forming a semisoft decay which renders the fruit worthless.

Causal Organism

Colletotrichum coccodes survives between crops on infested plant debris in the soil. Early in the growing season, spores from the soil splash on lower leaves of the tomato plant. Few symptoms develop on infected leaves, but the spores produced on foliage can be carried by splashing rain to developing green fruit. Infected green fruit will not develop symptoms of anthracnose until they begin to ripen. Ripe fruit is very susceptible to this fungus.

Control

- Rotate 3 years between pepper and tomato crops.
- Plant tomatoes and peppers in well drained fields to avoid excess soil moisture as fruit ripen.
- Apply overhead irrigation during the early part of the day so that plants dry before sundown.
- Harvest and use fruit before it fully ripens.
- If conditions favour development of anthracnose, a

preventative spray programme may be required to give adequate control of this disease. Apply registered fungicides according to product label instructions when weather conditions are above 65 degrees F and the foliage is likely to remain wet longer than 6 hours.

Applications to tomatoes should begin when the first fruit is larger than a walnut. Applications to peppers should be started as soon as fruit is present.

ROOT KNOT NEMATODE

Root knot first attracted attention in 1855 when it was found in a greenhouse in England. Since then it has been a cause for concern over much of the world. It affects more than 2,000 species of plants including economically important forage crops, small grains, fruits, vegetables, field crops, nursery crops, and turf grasses. Root knot nematodes are cosmopolitan in distribution and occur widely in the United States.

Symptoms

Root knot is very distinctive because of the galls or swellings produced on roots and underground portions of stems. These deformations can often completely ruin crops for sale. Plants, if infected when young, will be stunted, more susceptible to drought stress, and show symptoms of nutrient deficiency.

Large and small roots may be affected with swellings varying from round sphere-like galls to elongated spindles formed from large numbers of individual galls growing together. Root knot galls involve the entire root in the affected zone. They do not take the form of easily detached galls like those produced by nitrogen-fixing bacteria on the roots of legumes. Root knot galls should not be confused with the fungal disease, club root, on cruciferous crops.

In most instances, root knot is characterized by smaller swellings, and more uniformly distributed infection on the lateral feeding roots than is typically seen with clubroot. When the galls formed by the root knot nematode are broken open, shiny white bodies about the size of a pinhead, the enlarged female nematodes, are usually found. Also, glistening white to yellow egg masses are present on the root surface.

The galls formed by clubroot do not possess this characteristic and are usually pinkish or brick coloured. Phenoxy-type herbicides such as 2,4 D can cause swellings on stems of cruciferous crops which superficially look like root knot galls. Herbicide damage on these crops will not affect the roots, however. On young potato tubers the outer surface appears rough and warty because of the enlarged females underneath the skin.

Causal Organism

Root knot is caused by a small round worm, Meloidogyne spp. There are about 40 species at the present time. Not all plants are susceptible to any one species.

Juveniles emerging from eggs in the soil penetrate between and through cells to a position at the centre of the root usually near the growing tip. Feeding by the nematodes causes increases in root cell numbers and size. Enlarged cells, called giant cells, serve as food sources for female root knot nematodes.

During feeding, juveniles which will become females undergo a series of molts and enlarge. The female does not move from the feeding site. Each female deposits 300-500 eggs in a protective, jelly-like matrix at the root surface. Eggs, particularly in the egg mass, can withstand unfavourable environmental conditions. They represent one means of overwintering. Root knot nematodes may also overwinter in the soil in a juvenile stage. Root knot nematodes are often introduced into a growing area in infected planting stock.

Control

- Rotate at least 3 years with non-host crops. While the feeding habits of field populations must be determined before crop rotation plans are set, the most common root knot species will not reproduce on grasses, small grains, or corn. Clean cultivation to keep down weed hosts of the nematodes is important.

- Use transplants and nursery stock certified free of root knot nematode.
- Plant resistant varieties wherever possible.
- In commercial production, soil can be treated with fumigant-type nematicides.

PHYTOPHTHORA BLIGHT OF PEPPER AND CUCURBITS

Phytophthora blight, a highly destructive disease of peppers and cucurbits. It can become a serious problem during periods of heavy rainfall; the pathogen can spread rapidly through the crop, resulting in severe losses within a short time.

Symptoms

Peppers

Phytophthora blight affects both seedlings and mature plants. Infected seedlings show typical damping-off symptoms. Infection of older plants usually begins at or below the soil line. Water-soaked, dark brown lesions on the lower stems (collar rot phase) usually extend upward for an inch or more above the soil line and may expand to girdle the stems, preventing upward movement of water and nutrients.

This often results in a sudden wilting of foliage. Root infections may also occur which kill roots and cause wilting of the plant without the appearance of stem cankers. The foliar phase of this disease commonly occurs at forks in the branches, resulting in dark, girdling cankers and wilting of leaves and fruits.

Infected leaves develop circular or irregular, dark green, water-soaked lesions which dry and appear light tan. Fruit lesions may also appear as enlarging, watersoaked areas, which then shrivel and darken. A mass of white fungal growth may develop inside the fruit, and seeds usually turn dark brown or black.

A fine, grayish-white to tan mold may also become

evident over the lesion on the fruit surface. Under humid conditions, fungal growth develops extensively over the entire fruit.

Cucurbits

Seedlings show typical symptoms of damping-off. Mature plants may wilt suddenly as a result of root infection, in the absence of obvious stem or vine lesions. Stem and leaf petiole lesions are light to dark brown, watersoaked and irregular, eventually.

Becoming dried and papery. Fruit symptoms begin as small, water-soaked circular lesions, which expand to become large, soft, sunken areas covered with white fungal growth. Infected fruit often collapse or "melt down" in the field or in storage.

Causal Organisms

Phytophthora blight is caused principally by the soil-borne fungus, Phytophthora capsici, although a similar fungus, Phytophthora parasitica, has also been reported to cause fruit rot on peppers and cucurbits. Both of these fungi are dependent on free water in soil for infection. Because of this, initial infection by Phytophthora usually occurs on plants growing in poorly drained areas of fields.

The fungi that cause Phytophthora blight survive as thick-walled, resistant spores (oospores) in the soil and as mycelium in infected plant tissues. Once introduced, these fungi can survive up to 15 months in moist soil in the absence of host plants.

They can also be carried on seed or transplants. Once stem infection occurs, the fungi produce spores on infected stem tissues, which are then carried by splashed rain onto nearby plants.

Lower branches of adjacent plants can also be infected by rain-splashed soil contaminated by run-off water. Fruits in contact with the soil are especially prone to infection. Spores are produced on newly infected fruit and stems, and new infections can develop quickly in a short time. Warm (68 to

75 degrees F), wet weather is most favourable for disease development and spread of the pathogen. Disease progress declines when dry weather returns.

- Use only certified disease-free seed or transplants.
- Produce pepper plants on raised beds to retard initial stem infection. Arrange beds to prevent puddle formation between rows. Avoid depressions on the surface of the beds where water can accumulate, especially around the base of plants. Avoid planting peppers or cucurbits in poorly drained fields.
- Practice crop rotation so that peppers or cucurbits are grown only every 3 to 4 years to reduce the amount of Phytophthora present in the soil. Use non-susceptible crops such as corn, small grains, beans, crucifers, or potatoes in the rotation.
- Application of fungicides may reduce disease development and spread if used in combination with cultural practices such as crop rotation and raised beds. Application at planting and two or more (depending on the fungicide used) applications after planting are required.

BLOSSOM-END ROT OF TOMATO, PEPPER, AND EGGPLANT

Blossom-end rot is a serious disorder of tomato, pepper, and eggplant. Growers often are distressed to notice that a dry sunken decay has developed on the blossom end of many fruit, especially the first fruit of the season. This nonparasitic disorder can be very damaging, with losses of 50 per cent or more in some years.

Symptoms

On tomato and eggplant, blossom-end rot usually begins as a small water-soaked area at the blossom end of the fruit. This may appear while the fruit is green or during ripening. As the lesion develops, it enlarges, becomes sunken and turns black and leathery. In severe cases, it may completely cover the lower half of the fruit, becoming flat or concave.

Secondary pathogens commonly invade the lesion, often resulting in complete destruction of the infected fruit. On peppers, the affected area appears tan, and is sometimes mistaken for sunscald, which is white. Secondary molds often colonize the affected area, resulting in a dark brown or black appearance. Blossom end rot also occurs on the sides of the pepper fruit near the blossom end.

Cause

Blossom-end rot is not caused by a parasitic organism but is a physiologic disorder associated with a low concentration of calcium in the fruit. Calcium is required in relatively large concentrations for normal cell growth. When a rapidly growing fruit is deprived of necessary calcium, the tissues break down, leaving the characteristic dry, sunken lesion at the blossom end. Blossom-end rot is induced when demand.

For calcium exceeds supply. This may result from low calcium levels or high amounts of competitive cations in the soil, drought stress, or excessive soil moisture fluctuations which reduce uptake and movement of calcium into the plant, or rapid, vegetative growth due to excessive nitrogen fertilization.

- Maintain the soil pH around 6.5. Liming will supply calcium and will increase the ratio of calcium ions to other competitive ions in the soil.
- Use nitrate nitrogen as the fertilizer nitrogen source. Ammoniacal nitrogen may increase blossom-end rot as excess ammonium ions reduce calcium uptake. Avoid over-fertilization as side dressings during early fruiting, especially with ammoniacal forms of nitrogen.
- Avoid drought stress and wide fluctuations in soil moisture by using mulches and/or irrigation. Plants generally need about one inch of moisture per week from rain or irrigation for proper growth and development.
- Foliar applications of calcium, which are often

advocated, are of little value because of poor absorption and movement to fruit where it is needed.

CLUBROOT OF CRUCIFERS

Clubroot is a world-wide problem in temperate climates in the production of cruciferous vegetables such as cabbage, broccoli, cauliflower, radishes, and turnips; and field crops such as mustard and rape. The disease was known as early as the 13th century in England where it was called "finger and toe" disease because of the shape of infected roots.

Symptoms

The most striking symptom of clubroot is an abnormal enlargement of the root system with clubs often thickest at the centre, tapering spindle-like towards the ends. In radishes, clubroot causes distorted swellings on the base of the bulb and along the tap root.

In severe cases, entire plantings are destroyed. Clubroot-infected plants often wilt on sunny days and permanent wilting may accompany advanced decay of infected roots. Severe stunting may be evident if infection occurs early and the disease progresses rapidly. The malformed and greatly enlarged roots are the key symptom of this disease.

Causal Organism

Clubroot is caused by the soil-borne fungus, Plasmodiophora brassicae, which only infects plants in the crucifer family. It infects susceptible host plants through root hairs. Once in the tissue, it stimulates abnormal growth of affected parts, resulting in a swollen club. Infection is favoured by excess soil moisture and low pH, although it can occur over a wide range of conditions.

Once a plant is infected, numerous resistant spores of the fungus are produced in the "clubbed" tissues. As these tissues decay, spores are released into the soil where they can remain infectious for at least 10 years. Contaminated soil moved by wind or water can serve as a source of infestation of nearby fields causing outbreaks of disease in areas where susceptible crops are planted for the first time. Numerous races of the pathogen have been identified.

Clubroot is an very difficult disease to manage, and heavily infested areas may have to be abandoned for crucifer production. Some control may be achieved with the following measures:

- A good crop rotation programme, growing crucifers on the same soil no more than every third or fourth year, is essential to retard development of a large population of spores on land not already heavily infested.
- Since clubroot is favoured by a low pH, liming soil to pH 7.2 or above may be helpful. Raising soil pH too high, however, may interfere with the growth of succeeding crops other than crucifers.
  Calcitic lime is usually preferable to dolomitic lime, except for soils low in magnesium, where dolomitic lime is more effective. In course-textured soils, increasing the pH can result in boron deficiency. This may be alleviated by application of boron in transplant water or as a foliar spray.
- With transplanted crops, the use of pathogen-free seedbeds and uninfected plants is essential to prevent introduction of the disease.
- Application of a fungicide in transplant water or rototilled in a band prior to planting may help to reduce disease development.
- Clean and disinfect all machinery before moving it from infested to non-infested land.
- Some resistant cultivars are available. However, plant resistance has not been very useful in clubroot control because of rapid development of new races of the fungus.

COMMON SMUT OF CORN

Corn smut is an extremely common disease of sweet, pop, and dent corn and throughout the world. It is usually not economically important, although in some years yield losses in sweet corn may be as high as 20 per cent. In Mexico, immature smut galls are consumed as an edible delicacy known as cuitlacoche, and sweet corn smut galls have become a high value crop for some growers in the NE United States who sell them to Mexican restaurants.

Symptoms

The corn plant may be infected at any time in the early stages of growth, but becomes less susceptible after formation of the ear. Above-ground parts may be infected, but it is more common to see smut galls on the ears, tassels, and nodes than on the leaves, internodes, and aerial roots. The smut gall is composed of a great mass of black, greasy, or powdery spores enclosed by a smooth white covering of corn tissue.

The gall may be 4-5 inches in diameter. When leaves are infected, small pustules develop, usually on the midrib, causing some leaf distortion. After the spores mature, the outer covering becomes dry and brittle, breaks open, and the spores sift out. Greatest yield losses occur when the ear becomes infected or if smut galls form on the stalks immediately above the ears.

Causal Organism

Corn smut is caused by the fungus, Ustilago zeae, that survives as a resistant spore in the soil over winter, and possibly for 2 to 3 years. These spores, called teliospores, can be blown long distances with soil particles or carried into a new area on unshelled seed corn and in manure from animals that are fed infected corn stalks. The teliospores germinate in moist air and give rise to tiny spores called sporidia.

The sporidia bud like yeast, forming new spores that germinate in rain water that collects in the leaf sheaths. This leads to infections that are visible in 10 days or more. Wounds from various injuries provide points for the fungus to enter the plant.

The smut fungus is sensitive to temperature and moisture changes. In a warm season, the amount of smut is related closely to the amount of soil moisture, especially during June. When temperatures are lower than normal, there may be little smut even though soil moisture remains high.

- Seed treatment is of no value because few spores are on the corn seed.
- For the home garden, remove smut galls before they break open and bury or burn them. This must be done on a community basis, however, in order to be effective. Smut gall removal is not practical in commercial production.
- In order to reduce infection points from insect injury, control corn borers as first tassels appear by application of insecticides when insect populations are high.
- Avoid injury of roots, stalks, and leaves during cultivation.
- Deep plow diseased corn stalks in the fall to bury surviving spores.
- Use resistant hybrids or varieties. Dent corn is generally more resistant than sweet or popcorn. In sweet corn, the larger, later-growing varieties usually are more resistant than the smaller, early varieties.

Bacterial Spot, Speck, and Canker of Tomatoes

Bacterial spot, bacterial speck, and bacterial canker are widespread diseases of tomato that can cause localized epidemics during warm (spot and canker) or cool (speck), moist conditions. Bacterial spot can cause moderate to severe defoliation, blossom blight, and lesions on developing fruit. Bacterial speck also causes these symptoms but is usually not as severe as bacterial spot. Bacterial canker causes wilt, vascular discolouration, scorching of leaf margins, and lesions on fruit.

Symptoms

Foliar symptoms of bacterial spot and speck are identical. Small, water-soaked, greasy spots about 1/8 inch in diameter appear on infected leaflets. After a few days, these lesions are often surrounded by yellow halos and the centres dry out and frequently tear. Lesions may coalesce to form large, irregular dead spots. In mature plants, leaflet infection is most concentrated on fully-expanded and older leaves and some defoliation may occur.

Spots may also appear on seedling stems and fruit pedicels. In some cases, blossom blight may occur, causing flower abortion. This is more severe with bacterial spot and may result in a split fruit set which is especially troublesome with determinate cultivars intended for mechanical harvest.

Bacterial spot and speck can usually be differentiated by symptoms on immature fruits. Bacterial spot lesions are small water-soaked spots that become slightly raised and enlarged until they are about 1/4 inch in diameter. The centres of these spots later become irregular, light brown, slightly sunken with a rough, scabby surface. In the early stages of infection, a white halo may surround each lesion at which time it resembles the fruit spot of bacterial canker.

Small lesions which have not yet become scabby are often confused with lesions of bacterial speck. Bacterial speck appears on immature fruit as a black, slightly sunken stippling, eventually causing lesions less than 1/16 inch in diameter. Fruit lesions are not initiated on mature fruit in either disease. Primary or systemic symptoms of bacterial canker (from infections originating in seeds or young seedlings) include stunting, wilting, vascular discolouration, development of open stem cankers, and fruit lesions.

When affected stems are split open lengthwise, a thin, reddish-brown discolouration of the vascular tissue is observed, especially at the base of the plant. On young seedlings in the greenhouse, lesions may appear as raised pustules on leaves and stems.

These plants rarely survive the season in the field. Secondary symptoms in the field include leaf "firing" (necrotic marginal leaf tissue adjacent to a thin band of chlorotic tissue) and fruit lesions. Spots on fruit are relatively small (1/32 to 1/16 inch) surrounded by a white halo ("bird's-eye" spots). Canker bacteria may also invade internal fruit tissues, causing a yellow to brown breakdown.

Bacterial spot is caused by the bacterium, Xanthomonas campestris pv. vesicatoria, which can be carried as a contaminant on the surface of infested seed and has been found to overwinter in soil associated with plant debris. Bacterial speck is caused by another bacterium, Pseudomonas syringae pv. tomato.

This bacterium may also be seedborne and can overwinter on plant debris in soil and on the roots of many perennial plants. Bacterial canker is caused by Clavibacter michiganensis subsp. michiganensis, which, unlike the spot and speck pathogens, has the ability to infect tomato plants systemically. It is seedborne and can survive on infested plant debris in soil.

All three organisms may exist at low populations on leaf surfaces of symptomless plants. At the onset of favourable conditions, these low populations can increase rapidly and bacteria can then enter plants through stomata or small wounds and begin infection.

Bacteria can spread rapidly with spattering rain and widespread epidemics may develop. Penetration of tomato fruit occurs through wounds created by windblown sand, breaking of hairs, or by insect punctures. Optimal conditions for bacterial spot and canker are high moisture, high relative humidity and warm temperatures (75 to 90 degrees F). Bacterial speck is more likely to occur under cool (64 to 75 degrees F), moist conditions.

- Rotate tomatoes with non-solanaceous crops with at least 2 to 3 years between tomato crops. Avoid rotation with peppers, which are also susceptible to bacterial spot.
- Plant only seed from disease-free plants or seed treated to reduce any bacterial populations. Treatments include:
  - Fermentation of tomato pulp and seeds at room temperature for 4ñ5 days;
  - Soaking seeds in 0.6ñ0.8 per cent acetic acid for 24 hr at 70 degrees F;
  - Soaking seeds 5ñ10 hr in 5 per cent hydrochloric acid;

- Hot water treatment of seeds (122 degrees F for 25 minutes); or
- Sodium hypochlorite (bleach) treatment [20ñ40 minute soak of seeds in 1 per cent sodium hypochlorite (20 per cent bleach)]. Some decrease in germination may be expected from these treatments.

- Use only transplants free of disease symptoms.
- Carry out proper sanitation of transplant production greenhouses. Remove all weeds and plant debris, clean all tools with disinfectant solution, and wash hands thoroughly before and after handling plants. Water plants early in the day to reduce the amount of time foliage is wet.

  Do not handle plants when they are wet. After each crop, clean greenhouse walls, benches, etc., with hot soapy water, followed by thorough rinsing and treatment with a disinfectant.

  If possible, close up greenhouse after transplant production is completed to allow natural heating during the summer. Use only new plug trays and pathogen-free planting mixes. Avoid growing peppers and tomatoes in the same greenhouse unless pepper seed has also been treated as in step 2.
- In the field, control irrigation to minimize the time foliage is wet and avoid working among wet plants to minimize spread of disease.
- Applications of mancozeb plus copper soon after transplanting may help retard development and spread of bacterial spot and speck. This practice is not particularly effective for management of bacterial canker. Many tomato processors will not accept tomatoes treated with mancozeb or other EBDC fungicides. Check with your processor before applying one of these fungicides.

BACTERIAL WILT OF CUCURBITS

After vine crops begin to run, gardeners and farmers often

notice individual leaves with severe wilt symptoms on sunny days. Within a week or two the condition spreads to entire vines which do not recover from the wilt. This disease, called bacterial wilt, is especially common with cantaloupes and cucumbers. Squash and pumpkins may not wilt as rapidly, but may be dwarfed with extensive blossoming and branching. Watermelons are rarely affected.

Symptoms

Wilting of individual leaves or vines of the plant is the characteristic symptom. One or a few leaves wilt and become dull green. The disease spreads from the leaves downward into the petioles and then the stem until the entire plant wilts and dies.

There are other factors, such as vine borers and soil-borne fungal pathogens, that may cause cucurbits to wilt. Sometimes, if an affected stem is cut off near the ground, the sap may be milky in appearance or sticky and, if touched with the finger, the sap will string up to half an inch. This is a helpful test in diagnosis of bacterial wilt, but cannot be depended upon for positive identification.

Causal Organism

This disease is caused by a bacterium, Erwinia tracheiphila, that overwinters in the bodies of the striped and 12-spotted cucumber beetles. In the spring, the beetles emerge from the ground and feed on young plants, introducing bacteria into the leaves or stems.

The bacteria reproduce in the water-conducting vessels, producing gums that interfere with water transport. The beetles and bacteria are so intimately related that controlling the beetles will control infection by the bacteria. Once infection has occurred, however, no control is possible and wilting plants should be removed, if practical. The disease is not seed-borne.

Management

The only practical management measure is to use an

insecticide when seedlings first emerge to control the black and yellow cucumber beetles. Early infections are most severe, but total control depends on applications continuing at frequent intervals during the growing season. In some cases, if insect pressure is heavy, it may be necessary to apply an insecticide.

When plants are just cracking the soil, but have not yet emerged. Management of this disease is completely linked with preventing feeding of cucumber beetles on susceptible hosts.

FUSARIUM AND VERTICILLIUM WILTS OF TOMATO, POTATO, PEPPER, AND EGGPLANT

Solanaceous crop plants (tomato, potato, pepper, and egg- plant) may be infected at any age by the fungi that cause Fusarium wilt and Verticillium wilt. The wilt organisms usually enter the plant through young roots and then grow into and up the water conducting vessels of the roots and stem.

As the vessels are plugged and collapse, the water supply to the leaves is blocked. With a limited water supply, leaves begin to wilt on sunny days and recover at night. Wilting may first appear in the top of the plant or in the lower leaves. The process may continue until the entire plant is wilted, stunted, or dead.

Tomato and potato plants may recover somewhat but are usually weak, unthrifty, and produce fruit of low quality. Peppers typically collapse rapidly and die. Fusarium and Verticillium wilts are rarely significant in field grown tomatoes due to the widespread incorporation in tomato cultivars of genes for resistance to the pathogen.

However, the resurgent interest in planting "heirloom" tomato varieties which do not carry resistance genes has resulted in increased incidence of Fusarium and Verticillium wilts. Additionally, new races of both pathogens have been identified that are capable of overcoming the resistance in many popular tomato varieties.

Verticillium race 2 is now common in tomatoes, but its

importance in reducing yield is not known at this time. There is very little genetic resistance available to either disease in pepper or eggplant.

Symptoms

Fusarium Wilt

Fusarium wilt symptoms begin in tomato and potato as slight vein clearing on outer leaflets and drooping of leaf petioles. Later the lower leaves wilt, turn yellow and die and the entire plant may be killed, often before the plant reaches maturity. In many cases a single shoot wilts before the rest of the plant shows symptoms or one side of the plant is affected first. If the main stem is cut, dark, chocolate-brown streaks may be seen running lengthwise through the stem.

This discolouration often extends upward for some distance and is especially evident at the point where the petiole joins the stem. Potato tubers may show browning of the vascular ring as well as browning at the stem end and decay where stolons are attached. In pepper, lower leaves do not begin to wilt until roots and the base of the stem have already started to decay. Wilting of the entire plant soon follows. Dark brown, sunken, and eventually girdling cankers may be seen at the base of the pepper plant. In eggplant, wilting progresses from lower to upper leaves, followed by collapse of the plant.

Verticillium Wilt

Verticillium wilt symptoms on tomato, potato, and eggplant are similar to those of Fusarium wilt. Often no symptoms are seen until the plant is bearing heavily or a dry period occurs. The bottom leaves become pale, then tips and edges die and leaves finally die and drop off. V-shaped lesions at leaf tips are typical of Verticillium wilt of tomato. Infected plants usually survive the season but are somewhat stunted and both yields and fruits may be small depending on severity of attack.

A light tan discolouration in the stem similar to that caused by Fusarium wilt may be found but is usually confined to lower plant parts. The discolouration is typically lighter in colour than with Fusarium wilt. Symptoms on one side of the plant only are sometimes seen.

In potatoes the pathogen may be part of a complex that includes, among others, the root lesion nematode and the bacterial soft rot organism, resulting in premature plant death ("potato early dying disease"). Tubers from Verticillium-infected plants may show light brown vascular discolouration, usually restricted to the stem end. In pepper, the lower leaves wilt, then leaf tips and margins dry and turn brown. Brown streaks in the vascular tissue can be observed well up into the plant, which rapidly collapses and dies.

Causal Organisms

Fusarium wilt in solanaceous crops is caused by several different types of the fungus Fusarium oxysporum. These are: F. oxysporum f. sp. lycopersici (tomato), F. oxysporum f. sp. melongenae (eggplant) and F. oxysporum var. vasinfectum (pepper). Fusarium wilt in potato is caused by a complex of up to four different Fusarium spp. All of the Fusarium wilt pathogens are generally specific to their hosts and are soilborne. They are warm weather organisms, and therefore Fusarium wilts are most serious later in the growing season.

Verticillium wilt is caused by the fungi Verticillium albo-atrum and V. dahliae. These fungi attack a wide range of plant species, including cultivated crops and weeds. They are soilborne in field and greenhouse soils where they can persist for many years. V. albo-atrum is a cool weather organism that grows best when soil temperatures are between 65 and 75 degrees F. V. dahliae is more active between 75 and 83 degrees F.

Although disease is retarded by the higher temperatures that favour Fusarium wilt, visible symptoms may appear to be more severe when high temperatures exist, due to restricted water movement in the plant brought about by damage done to the water conducting vessels earlier in the growing season.

- Because Fusarium and Verticillium fungi are

widespread and persist several years in soil, a long crop rotation (4 to 6 years) is necessary to reduce populations of these fungi.

Avoid using any solanaceous crop (potato, tomato, pepper, eggplant) in the rotation, and if Verticillium wilt is a problem, also avoid the use of strawberries and raspberries, which are highly susceptible. Rotate with cereals and grasses wherever possible.

- Keep rotational crops weed-free (there are many weeds hosts of Verticillium).
- Whenever practical, remove and destroy infested plant material after harvest.
- Maintain a high level of plant vigour with appropriate fertilization and irrigation, but do not over-irrigate, especially early in the season.
- Plant disease resistant tomato varieties, labeled V (for Verticillium) and F (for Fusarium). These disease resistance designations are usually shown in seed catalogues. Fusarium- or Verticillium-resistant varieties of eggplant, potato, and pepper are generally not available.
- If soils are severely infested, production of solanaceous crops may not be possible unless soil fumigation is an option.

BACTERIAL SPOT OF PEPPER

Bacterial spot of pepper is one of the most destructive diseases of pepper in climates where high temperature and frequent rainfall occur during the growing season. The disease causes spots on leaves and fruit; leaf defoliation; and a reduction in plant growth, fruit yield, and quality. Bacterial spot is also a serious problem on tomato, although not all strains of the pathogen can cause disease on both hosts.

Symptoms

Characteristic bacterial spot symptoms can appear on the leaves, fruits, stem, and petioles. On leaves, symptoms begin as small, yellow-green circular lesions surrounded by a yellowish halo. These spots appear water-soaked under wet conditions. As the lesions mature, a general yellowing extending from the area around the lesions develops on diseased leaves and the centre of the spots become brown to black and sunken.

Tissue in the centre of the lesion often dries and breaks away, giving a "shot-hole" appearance to the leaf. When spots are numerous, they may join together and form irregular discoloured streaks along the veins and leaf margins. Edges and tips of leaves may die, then dry and break away, causing leaves to appear ragged.

Severely spotted leaves turn yellow or brown and fall from the plant; young leaves can be distorted. Fruit spots begin as green, circular, slightly raised lesions which eventually become brown or dark, raised, and about 1/8 inch in diameter. Centres of the spots become necrotic, corky, and scab-like.

On stems and petioles, lesions are elongated and blackened, and can kill leaflets. Cotyledons are particularly susceptible to bacterial spot; lesions are initially small, sunken, and silvery. They later become darker in colour.

Causal Organism

Bacterial spot of pepper is caused by a bacterium, Xanthomonas campestris pv. vesicatoria, which can overwinter in crop residue in or on soil, on or in seeds, and on wild host plants. Pepper seeds infested with the pathogen are a major source of inoculum for bacterial spot as well as the major means of long distance spread of the pathogen. The pathogen can survive on dried seeds for up to 10 years.

It is not able to survive free in soil for a long time, but can survive up to 6 months in infected crop debris in soil. The bacterium penetrates leaves through stomata and/or wounds, and fruits through wounds created by wind-driven sand, insect punctures, or mechanical injury. Dissemination of the bacterium occurs between and within fields by water-splashing, aerosols, or during cultivating, hoeing, thinning of direct seeded plants, transplanting, or harvesting. In the

United States, numerous physiological races of the bacterium are known, most of which have been found.

- Use pathogen-free seeds and transplants.
- Use sodium hypochlorite-treated seed to reduce bacterial populations.
- Practice crop rotation with non-host plants such as corn and soybean so that peppers are grown only every 3 to 4 years. However, do not use soybeans in the rotation if white mold (Sclerotinia sclerotiorum) has been a problem.
- Deep plow to bury infected crop debris.
- Avoid working in the field when foliage is wet.
- Eliminate wild host plants such as nightshade and ground cherry in and around field.
- Application of copper-containing pesticides may be helpful for preventing development and spread of bacterial spot.
- None of the currently available pepper varieties are resistant to all known races of the bacterial spot pathogen. However, use of varieties resistant to one or more races of the pathogen may provide some control, depending on the races present.

DOWNY MILDEW OF CRUCIFERS

Downy mildew affects all cultivated plants and weeds in the crucifer family. It can be a serious problem in commercial production of cabbage, broccoli, cauliflower, radish, turnip, mustard, collard, and cruciferous greens. Under favourable conditions, it may cause serious losses in the field or may develop after harvest and cause deterioration of product quality during packing and shipping.

Symptoms

Plants can be infected at any stage of development. In seed beds, cotyledons and primary leaves are invaded resulting in fungal growth visible on the underside of the leaf. Later a slight yellowing develops opposite the fungal growth on the upper side of the leaf. The young leaf or cotyledon, when yellow, may drop off. Older leaves usually persist and infected areas gradually enlarge, turn bright yellow, then become tan and papery. Rarely the affected leaf may develop hundreds of minute darkened specks. Under cool, moist conditions, a white mildew growth can be seen on the underside of infected leaf lesions.

Symptoms may appear on other plant parts as well. The fleshy roots of turnips and radishes may develop an internal, irregularly shaped discolouration extending from the crown downward. The flesh may be brown or black or show a form of net necrosis. In advanced stages, the skin becomes roughened by minute cracks and the root may split open. In radish these symptoms may be confused with those caused by Rhizoctonia.

Causal Organism

The fungus causing downy mildew in crucifers, Peronospora parasitica, overwinters in roots or in decaying portions of diseased plants. Thick-walled resting spores may form in stems, cotyledons, and other fleshy parts of infected host plants. On growing plants, the fungus produces large numbers of spores that are blown about by wind and splashed by rain. Moisture and temperature are important in the spread and reproduction of this fungus.

High relative humidity during cool or warm, but not hot, periods promotes its growth and sporulation. Presence of a water film on the foliage from fog, drizzling rain, or dew allows spores to germinate, infect, and produce more spores on a susceptible host in as few as 4 days.

- Use a crop rotation plan that excludes production of any type of cruciferous crop for at least 2 out of every 3 years.
- Practice sanitary measures such as the use of clean seed beds away from other crucifer production and the destruction of cruciferous weeds.
- Use a planting site and plant spacing pattern that expose plants to full sun throughout the day.
- If severe disease pressure is expected, apply a

registered fungicide weekly beginning soon after emergence.

- Disease resistant cultivars are not available for most cruciferous crops. However, some hybrid cultivars of broccoli are resistant or tolerant to downy mildew.

BLACK ROT OF CRUCIFERS

Black rot is the most serious disease of crucifer crops world wide when environmental conditions (relatively high temperature and humidity) are favourable. The disease affects primarily aboveground parts of plants at any stage of growth and causes high yield and quality losses, especially in tropical and subtropical regions during the rainy season. All vegetables in the crucifer family, including broccoli, Brussels sprouts, cabbage, cauliflower, Chinese cabbage, kale, mustard, radish, rutabaga, and turnip, are susceptible to black rot. Many cruciferous weeds such as Shepherd's Purse, wild mustard, and yellow rocket may also be hosts of this pathogen.

The characteristic black rot symptom on most cultivated crucifer plants is the appearance of yellow, V-shaped lesions along the margins of leaves. The point of the V-shaped lesion is directed toward a vein.

When lesions enlarge, wilted tissue expands toward the base of leaves. Eventually the diseased areas become necrotic and the veins turn black or brown. The infection may move down the vascular tissue of petioles and then spread up and down the stems. When stems and petioles of an infected plant are cut crosswise or lengthwise, the black-brown vascular tissue with yellowish bacterial slime is observed. These symptoms may be confused with Fusarium yellows, except that Fusarium causes brown vein discolouration without bacterial slime.

Moreover, symptoms of black rot may vary according to age of host, host genus, species, and cultivar and even environmental conditions. For example, symptoms on cauliflower may appear as numerous black or brown specks, scratched leaf margins, black veins, and discoloured curds.

Many cruciferous weed species do not exhibit any of these characteristic symptoms even when infected.

Causal Organism

Black rot of crucifer is caused by a bacterium, Xanthomonas campestris pv. campestris (Xcc). The bacteria can overwinter in plant debris, in and on seeds from diseased plants, and in and on weeds. The pathogen may survive in diseased crop residue buried in soil for up to 2 years, but not more than 60 days free in soil. The major source of these bacteria is infected seeds, which enable long-distance spread of the disease.

The pathogen is spread within and between fields by splashing water, wind, insects, machinery, and irrigation or drainage waters. The bacteria infect the cotyledons and young leaves through natural plant openings (stomata, hydathodes) or wounds and then migrate between cells until they reach the xylem tissue where they spread throughout the plant. Free moisture is required for infection by the pathogen. After infection, symptoms may appear on plants within 7 to 14 days under optimum conditions (25 to 30 degrees C).

Effective management of black rot of crucifers depends on the application of the following practices in combination:

- Use black rot-tested, disease-free seed grown in an arid production area.
  - If source of the seeds is unknown, or infested seedlots must be used, treat seed with hot water to eradicate pathogenic bacteria. Cabbage, broccoli, and Brussels sprouts can be treated at 50 degrees C for 25 minutes, while seeds of cauliflower, kale, turnip, and rutabaga are treated for 15 minutes. However, this treatment may reduce the viability of seed.

    Therefore, some other chemical seed treatments, including, sodium hypochlorite, hydrogen peroxide, and hot acidified cupric acetate or zinc sulfate can be applied to eliminate the bacteria from crucifer plant seeds.

- Use certified disease-free transplants.
- Practice crop rotation where crucifers are grown only every 3 to 4 years to eliminate the inoculum sources from diseased crop debris in the soil.
- Good sanitation practices should be performed to prevent disease spread.
  - Eliminate all volunteer crucifer plants from previous crops and alternative wild host plants within and around the field.
  - Do not apply manure that may contain crucifer residues.
  - Do not use sprinkler irrigation.
  - Avoid working in the field when plants are wet.
  - Do not allow machinery and equipment movement from infested areas to non-infested fields.
  - Deep plow to bury all crucifer residues after harvest.
- Application of fixed copper pesticides in the field may help to reduce spread of the disease.
- A few black rot-resistant cultivars of cabbage and other crucifers are commercially available. These resistant cultivars should be used in crucifer growing regions where black rot is a common problem.

GUMMY STEM BLIGHT AND BLACK ROT OF CUCURBITS

Gummy stem blight is an important disease of squash, pumpkins, cucumbers, watermelons and other field-grown cucurbit crops. It can occur at any growth stage, from seedlings to mature plants. This disease on fruits, in the field or in storage, is called black rot. The disease also can cause extensive damage to all above-ground parts of greenhouse-grown cucumbers.

Symptoms

Gummy stem blight occurs on all plant parts except roots. Leaf symptoms appear as dark yellow or reddish-brown lesions in various shapes. Lesions begin at leaf margins and extend rapidly back into the leaf blade, causing curling, shriveling, and death of the entire leaf. Pimple-like structures (pycnidia) may be found in leaf lesions by close inspection with a hand lens.

Fruit symptoms vary among crops. Winter squash (Hubbard, butternut, etc.) are likely to show symptoms primarily on fruit or older leaves. Black rot symptoms appear as a brown to black rot of the rind, flesh, and seed cavity accompanied by heavy white and black fungus growth.

Lesions may develop anywhere on the fruit, first as water-soaked areas dotted with pycnidia that ooze yellowish masses of spores. On Hubbard-type squash, the brownish-black rot extends down into the flesh and seed cavity. Seeds often become dotted with small black pycnidia. On butternut squash, lesions are brown with irregular ring patterns and are superficial over the skin surface, not penetrating into the flesh or seed cavity.

Infection usually occurs in the field, causing water-soaked, cracked, brownish cankers on the vines. A reddish gum may develop in these cracks, although this alone is not a diagnostic sign. (Fusarium and scab may also produce a reddish gum.) Fruit may decay at the site of attachment as a result of the fungus invading the stem.

Butternut squash and gourds may develop black rot before harvest, but Hubbard squash are resistant during growth and do not show symptoms until the storage period. Fruit rot on greenhouse cucumbers usually begins at the blossom end of immature fruit. Lesions are firm and become dark brown to black when cut open. Occasionally, lesions develop on one side of a fruit, causing it to hook as it grows.

Causal Organism

Gummy stem blight is caused by the fungus Didymella bryoniae. The pathogen can be seed-borne or can survive on organic debris from previously infected cucurbits or on wild or volunteer cucurbits. The gummy stem blight fungus produces two types of spores. Windblown ascospores are likely to start the disease in a field. Later, pycnidiospores are released in a gummy substance that makes them adaptable for short distance spread by splashing water. Spore production and infection are influenced by moisture and temperature. A moisture film from dew, rain, or overhead irrigation is necessary for spore germination.

Optimum temperature for infection is 61 to 75 degrees F. Low night temperatures, particularly in greenhouses, may cause water droplets to exude from leaf points and condensation to form on leaves, favouring infection by the fungus at those points. Infection of fruits commonly occurs through harvest wounds or through dying flowers.

- Use only disease-free seed produced in arid western locations.
- Plow crop refuse deeply to reduce survival of the fungus as soon as crop is harvested.
- Practice crop rotation with non-cucurbit crops so that cucurbits are grown only every 3 to 4 years.
- Apply foliar-protectant fungicides on a routine basis.
- Avoid wounding fruit during harvest and store fruit at 45 to 50 degrees F to prevent postharvest fruit rots.
- Resistant cultivars are not currently available.

MUMMY BERRY OF BLUEBERRY

Mummy berry is one of the most serious diseases of blueberry. Once the disease becomes established in a planting, it can destroy most of the crop. Losses result from:

- Rotted berries; and
- Killing or blighting of blossoms, blossom and leaf clusters and young shoots.

Symptoms

Blighting of new shoot tips and blossoms can be easily mistaken for frost damage. By blossom time, the infected young leaf and shoot growth will wilt, turn brown and die. About a week or so after infection of the early new growth occurs, dead areas develop on the petioles and along the midrib and veins of the leaves or at the base of flowers. Berries that develop from infected flowers may attain nearly full size before turning tan or gray and shriveling into hard mummies, which drop to the ground at or before harvest.

Causal Organism

Mummy berry is caused by the fungus, *Monilinia vaccinii-corymbosi*. The fungus overwinters in the shriveled mummies on the ground. In early spring, cup- or globe-shaped structures of the fungus called apothecia are produced on mummified berries during cool rainy periods. Spores produced inside apothecia are released into the air and carried by the wind to young developing leaf shoots and flowers where they cause primary infections.

If moisture is not present, the fungus will not produce spores. However, the fungus may survive in mummies for one year or more. Another type of fungus spores (conidia) is produced on dead tissue that results from primary infections. These conidia are spread during bloom by wind and insects, and result in secondary infection of flowers. Fruit that develops from infected flowers turns into mummies and falls to the ground.

Control

Removing mummified berries from the planting will greatly aid in controlling the disease. Removing these berries is not practical on a commercial scale, but may be of value in backyard plantings where there are just a few plants. After removing mummies, burn or bury them.

Cultivation in early spring to disturb or cover the mummies has been reported to be effective. Mummies that are disturbed or covered with soil at this time remain dormant or do not produce spores. Cultivation between rows and raking under plants to disturb or cover mummies should be done as early as possible in the spring and repeated after each hard rain until after bloom. If just a few mummies are missed, they can produce enough spores to infect the planting.

Where mummy berry is a problem, a good fungicide spray programme is essential.

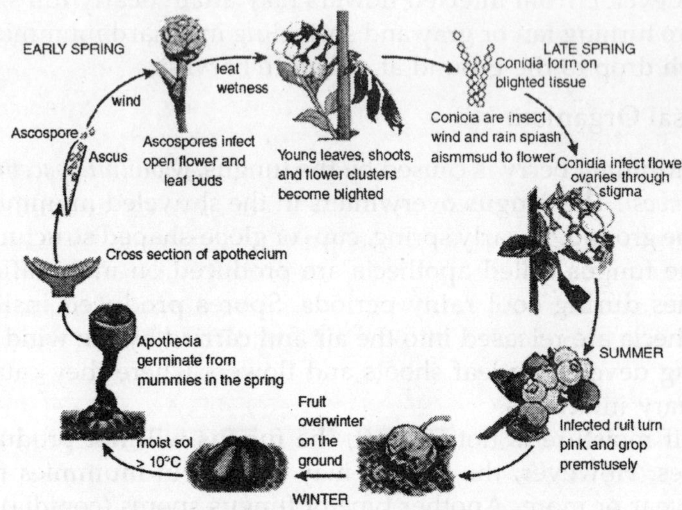

Fig. Disease cycle of mummy berry

LEATHER ROT OF STRAWBERRY

Leather rot of strawberry has been reported in many regions of the United States. In many areas, it is considered a minor disease of little economic importance. However, excessive rainfall during May, June, and July can lead to severe losses in fruit yield and quality resulting from leather rot. Commercial growers have lost up to 50 per cent of their crop to leather rot. The leather rot fungus primarily attacks the fruit, but many also infect blossoms.

Leather rot can infect berries at any stage of development. Where the disease has been a problem, infection of green fruit is common. On green berries, diseased areas may be dark brown or natural green outlined by a brown margin. As the rot spreads, the entire berry becomes brown, maintains a rough texture, and appears leathery. The disease is more difficult to detect on ripe fruit. On fully mature berries, infection may result in little colour change or discolouration ranging from brown to dark purple.

Infected ripe fruit are usually softer to the touch than healthy fruit. When diseased berries are cut crosswise, a marked darkening of the water-conducting system to each seed can be observed. In later stages of decay, mature fruits also become tough and leathery. Occasionally, a white moldy growth can be observed on the surface of infected fruit. In time, infected fruit dry up to form stiff, shriveled mummies.

Berries that are affected by leather rot have a distinctive, unpleasant odor and taste. Even the healthy tissue on a slightly rotted berry is bitter. This presents a special problem to growers in pick-your-own operations.

An infected mature berry with little colour change may appear normal and be picked and processed with healthy berries. Consumers have complained of bitter tasting jam and jelly made with berries from fields where leather rot was a problem. Leather rot is observed most commonly in poorly drained areas where there is or has been free standing water, or on berries in direct contact with the soil.

Causal Organism

Leather rot is caused by the fungus, *Phytophthora cactorum*. The fungus survives the winter as thick-walled, resting spores, called oospores, that form within infected fruit as they mummify. These oospores can remain viable in soil for long periods of time. In the spring, oospores germinate in the presence of free water to produce a structure called a sporangium.

A second type of spore, called a zoospore, is produced inside the sporangium. Up to 50 zoospores may be produced inside one sporangium. The zoospores have tiny flagella and can swim in a film of water. In the presence of free water on the fruit surface, the zoospores germinate forming a germ tube and infect the fruit.

In later stages of infection, sporangia are produced on the surface of infected fruit under moist conditions. The fungus is spread by splashing or windblown water from rain or overhead irrigation. Sporangia and/or zoospores are carried in water from the surface of infected fruit to healthy fruit where new infections occur. Under the proper environmental conditions, the disease can spread very quickly.

A wetness period (free water on fruit surface) of one hour is sufficient for infection. The optimum temperatures for infection are between 62 and 77 degrees F (17-25 degrees C). As the length of the wetness period increases, the temperature range at which infection can occur becomes much broader.

As infected fruit dry up and mummify, they fall to the ground and lie at or slightly below the soil surface. Oospores formed within the mummified fruit enable the fungus to survive the winter and cause new infections the following year, thus completing the disease cycle.

Control

- Select a planting site with good soil drainage and air circulation. Good soil drainage is critical. Sites that drain poorly or are subject to periodic flooding are ideal for leather rot development.
- Plants should be exposed to direct sunlight (avoid shade). Plant rows with the direction of the prevailing wind to promote faster drying of foliage and fruit.
- Mulch strawberry plants with straw or other material that reduces fruit contact with soil. Research has shown that a good layer of straw mulch is very beneficial in controlling leather rot.
- Proper spacing of plants and timing of fertilizer applications are important. Excessive applications of nitrogen fertilizer can produce excessive amounts of dense foliage.

  Shading of berries by thick foliage prevents rapid drying of fruit during wet periods and creates ideal conditions for disease development.
- Pick fruit frequently and early in the day (as soon as plants are dry). Cull out all diseased berries, but do not leave them in the field.
- When combined with the above mentioned cultural practices, fungicides can be beneficial in control. Fungicide use for controlling leather rot is generally not recommended in backyard fruit plantings. Homeowners are encouraged to use the previously

described cultural practices to eliminate the need for fungicides.

CANE BLIGHT OF RASPBERRIES

Cane blight is one of the more damaging diseases of raspberries. The disease is most common on black raspberries but also occurs on red and purple varieties. The disease occasionally occurs on blackberries and dewberries. Cane blight can result in wilt and death of lateral shoots, a general weakening of the cane, and reduced yield. It is usually most severe during wet growing seasons.

Symptoms

On first year canes (primocanes) dark brown-to-purplish cankers form on new canes near the end of the season where pruning, insect, and other wounds are present. The cankers enlarge and extend down the cane or encircle it, causing lateral shoots above the diseased area to wilt and eventually die. Black specks, which are reproductive bodies of the cane blight fungus, develop in the brown cankered bark.

In wet weather, large numbers of microscopic spores ooze out of the pycnidia. This ooze gives the bark a dark-gray, smudgy appearance. During winter, infected canes commonly become cracked, brittle, and snap off easily. On infected second-year canes (floricanes), the side branches may suddenly wilt and die, usually between blossoming and fruit ripening. Upon close examination, the presence of dark brown or purplish cankers can be observed on the main cane or branches below the wilted area.

Causal Organism

Cane blight is caused by the fungus, *Leptospaeria coniothyrium*. The pathogen survives over winter on infected or dead canes. The following spring, spores are released and carried by splashing rain and wind to nearby primocanes. Under moist conditions, the spores germinate and penetrate pruning wounds, insect punctures, fruit stem breaks and other wounds. After entry the fungus rapidly invades and kills bark and other cane tissues. Fungal fruiting bodies are formed in older cankers and complete the disease cycle. Dead canes continue to produce conidia and remain a source of infection for several years.

Control

- All steps possible should be taken to improve air circulation within a planting, to allow faster drying of foliage and canes. Reducing the number and duration of wet periods should reduce the potential for infection. Excessive applications of fertilizer (especially nitrogen) should be avoided, since it promotes excessive growth of very susceptible succulent plant tissue.

  Plants should be maintained in narrow rows and thinned to improve air circulation and allow better light penetration. Weeds are very effective in reducing air movement; therefore, good weed control within and between rows is important for improving air circulation within the planting.

  Raspberries should be planted in sunny, open areas where water and air drainage are good. This allows plants to dry quicker after wet periods, and reduces the chance of infection.
- Wild brambles, especially wild raspberries, growing in the area should be removed. They can provide a continuous source of spores to spread this and other diseases and pests to cultivated raspberries and blackberries.
- Healthy, rapidly growing plants that have been properly fertilized and watered, are more resistant to cane blight.
- After harvest, remove and destroy all old fruited floricanes and any new primocanes canes that are infected. Old canes should be removed before growth starts in the spring.
- Keep plantings free of insects, since they may cause wounds that serve as entry points for the fungus.

Avoid any other pests or cultural practices that result in wounding of the canes.

- If cane blight is a serious problem, the use of fungicides should be considered.

EUTYPA DIEBACK OF GRAPE

For more than 60 years the eastern grape industry recognized a disease called "dead-arm" thought to be caused by the fungus.

Phomopsis viticola. In 1976, researchers demonstrated that the "dead-arm" disease was actually two different diseases that often occur simultaneously: Eutypa dieback and Phomopsis cane and leaf spot. Eutypa dieback is the new name for the trunk and arm phase of what was once known as "dead-arm."

Scientists now propose that the name "dead-arm" be dropped. It is important that Eutypa dieback and Phomopsis of the main trunk. These cankers are usually difficult to see because they are covered with bark. One indication of a canker is a flattened area on the trunk. Removal of bark over the canker reveals a sharply defined region of darkened or discoloured wood.

These cankers may be up to 3 feet long, but do not enter one- or two-year wood and seldom go below ground line. When the trunk is cut in cross-section, the canker appears as darkened or discoloured wood extending in a wedge shape to the centre of the trunk.

The most striking and obvious symptoms of Eutypa dieback are the leaf and shoot symptoms, which may not develop for up to 3 years after infection of the vine has occurred. These symptoms are most obvious in spring (May and June). Spring shoot growth is weak and stunted above the cankered area. Leaves are at first smaller than normal, cupped, misshapen and yellowed. Later in the season (mid July), these leaf and shoot symptoms may disappear from all but the basal leaves of affected shoots. The vines may appear to have recovered. However, the infected trunk and all growth above it will eventually die.

CAUSAL ORGANISM

Eutypa dieback is caused by the fungus, *Eutypa armeniacea*. The fungus survives in infected trunks for long periods of time, whether they remain as part of the in-place vine or as prunings in the vineyard. Eventually, the fungus produces reproductive structures (perithelia) on the surface of infected wood. Spores (ascospore) are produced in these structures and are discharged into the air.

Ascospore discharge is initiated by the presence of free water (either rainfall or snow melt). Most spores appear to be released during winter or early spring, with relatively few being released during the summer months. Unfortunately, most spores are released at the same time pruning is being conducted.

The ascospore can be carried considerable distances by air currents to recent wounds on the trunk. Pruning wounds are by far the most important points of infection. When the ascospore come in contact with newly cut wood, they germinate and a new infection is initiated.

Control

- At present the primary control methods is removing infected trunks from the vineyard. The vine must be cut off below the cankered or discoloured wood. If the canker extends below the soil line, the stump can be left and a new trunk formed. It is important to remember that the best time to identify infected vines is spring.

  (May and June) and when the leaf and shoot symptoms are most obvious. If trunks cannot be removed in the spring, they should at least be marked so they can be removed after harvest, but before the next spring.
- Sanitation is critical. All wood from infected plants must be removed from the vineyard and destroyed (either buried or burned). An old infected trunk lying on the ground may continue to produce spores for several years.

- At present, no fungicide recommendations are available for control of this disease.

NECROTIC LEAF BLOTCH OF GOLDEN DELICIOUS APPLES

Necrotic leaf blotch is a common but minor disease that appears to be restricted to the Golden Delicious cultivar of apples and its sports. The disease is a physiological dis-order and has been observed for about 20 years in most apple-growing areas of the Eastern and Midwestern United States.

Necrotic leaf blotch usually appears in late June or early July and is most severe in the latter part of the growing season. Mature leaves from the base to the centre of upright, succulent, rapid-growing shoots are usually the only ones that show symptoms. Young, immature leaves on succulent shoots and mature or cluster leaves on fruiting wood are usually not affected.

The effect of the disease varies among individual Golden Delicious trees, both from orchard to orchard and from tree to tree within the same orchard.

Symptoms

The symptoms are a leaf blotching, followed by a yellowing of the affected leaves (chlorosis) and defoliation. Defoliation can be severe. Irregular brown blotches, usually 0.5 to 1.5 centimeters in diameter, suddenly appear on leaves. The blotches are usually restricted by the larger leaf veins. Many affected leaves turn yellow and drop all at once a few days after the necrotic blotches first appear. The most conspicuous symptom of necrotic leaf blotch is the large number of yellow leaves, on the tree and later on the ground, that suddenly appear during July and August.

A distinctive characteristic of the disease is its appearance in two to four, or more, waves of these symptoms during certain periods that are scattered throughout the latter half of the growing season. Between these periods or waves, little or no additional disease develops. During the course of the season, 10 to 50 per cent defoliation may occur on severely affected trees. All sports of Golden Delicious on any root stock appear to be equally susceptible.

Cause

Necrotic leaf blotch of Golden Delicious apples appears not to be caused by a fungus, bacterium, air pollutant, or a nutrient deficiency. The causal agent of necrotic leaf blotch is unknown; therefore, the disease is often referred to as a physiological disorder. The disease commonly appears during hot, hazy weather following a humid, rainy period. Other environmental factors or an air pollutant may be involved. The levels of ozone or sulfur dioxide in the air and foliar nutrients in the leaves have little or no effect on disease development.

Control

- It has been observed that certain fungicides result in less necrotic leaf blotch. Since a fungal pathogen does not appear to cause necrotic leaf blotch, it is difficult to assess the role of fungicides in reducing necrotic leaf blotch. The effect may be indirect, by altering leaf physiology. Fungicides containing zinc (Zn) ions appear to be the most effective. Golden Delicious trees sprayed, during the summer cover spray period, with fungicides containing mancozeb (Dithane M-45, Manzate D, Penncozeb) or Ziram tend to have less necrotic leaf blotch than unsprayed trees or trees sprayed with other fungicides.
- Trees that have an annual moderate crop of fruit have less necrotic leaf blotch than trees that have a biennial bearing habit or a light fruit crop.
- Golden Delicious trees pruned to a central leader commonly have less necrotic leaf blotch than trees pruned to an open centre.

WHITE PINE BLISTER RUST ON CURRANTS AND GOOSEBERRIES

White pine blister rust is not a serious disease of currants and gooseberries; however, it is a very serious disease of white pines (*Pinus strobus*). Currants and gooseberries serve as an alternate host for the rust fungus that causes white pine blister rust. Therefore, planting currants and gooseberries in areas where white pines are present can lead to serious losses of white pines.

North American white pine species, including bristlecone, limber, sugar, eastern white, southwestern white, western white, and whitebark, are highly susceptible. White pine blister rust causes significant damage in pine forests by forming cankers on the branches of white pines. These cankers ultimately kill the trees. Black currant is the most susceptible of the *Ribes* species.

To protect white pine forests, several states have enacted laws concerning planting of black currants.

The current law to suppress and control White Pine Blister Rust Disease is as follows:

- The European black currant, *Ribes nigrum* L. or any variety of this species is hereby declared to be a public nuisance, and it shall be unlawful for any person to possess, transport, plant, propagate, sell, or offer for sale, plants, roots, scions, seeds, or cuttings of these plants in this state.
- Recognized varieties, e.g., "Consort" produced by the hybridization of *Ribes nigrum* L. or a variety thereof with a resistant or immune species, known to be immune or highly resistant to the White Pine Blister Rust fungus, (*Cronartium ribicola,* Fischer) are exempt from the restrictions imposed by paragraph (A) above.

Symptoms

On *Ribes* in the spring, tiny yellowish spots become visible on the upper surface of the leaves, while on the underside, orange-yellow blister-like fruiting bodies appear. By late summer, yellow to brown threadlike growths develop on or near these infection spots on the leaf. Bushes also will have premature defoliation.

On white pine, the symptoms include dead branches,

chlorotic foliage, branch girdling by lesions that exude resin or sticky yellowish fluid (spermagonia), cankers that are diamond-shaped to elliptic with a dead centre surrounded by a band of yellowish-green infected bark, light yellow-orange aecia, and death of the tree.

Causal Organism and Disease Development

White pine blister rust is caused by a fungus, *Cronartium ribicola*. The organism was introduced from Europe in the early 20th Century. It has spread throughout the entire range of white pines in North America. The life cycle takes three to six years to complete.

The initial infection of black currant bushes occurs in the spring, when aeciospores from diseased white pine land on the leaves of the bush. These spores can travel on the wind several miles. Moisture is needed for the germination of the aeciospores.

After one to three weeks incubation, the plants begin to show the first symptoms of disease. Yellowish spots appear on the top side of the leaves and fruiting bodies (uredia) appear on the underside of the leaves. The fruiting bodies (uredia) produce uredospores which reinfect *Ribes* leaves. This cycle can be repeated many times during a single growing season.

In late summer or early fall, when day length and temperature decrease, telia (another type of fruiting body) replace the uredia on the underside of the leaves. The telia produce teliospores, which germinate in place at night during wet weather and produce basidiospores.

After telial formation, at least 60 hours of wet weather with temperatures not exceeding 20C are required for basidiospore formation, dispersal, and infection of white pine. These basidiospores are carried by the air currents to white pine trees. The basidiospores germinate on wet needles and enter through the stomata.

The only symptom initially is a yellow to reddish spot at the site of infection. Infected needles turn yellow and drop prematurely, but often not before the fungus has grown down the needle and entered the twig. This occurs late in the fall or early in the spring of the next year. About 12 to 18 months after infection, the fungus has grown from the needle into the bark of the stem, where it produces spermagonia.

During the first or second year of infection, the spermagonia produce aecia. The aecia produces powdery bright yellow aecidiospores that may live for many months. The aecidiospores are carried by the wind to Ribes bushes, completing the cycle.

Control

- Remove susceptible *Ribes* species and infected plants.
- Plant only disease-free resistant varieties of *Ribes* approved by the Department of Agriculture. Some examples of resistant varie-ties of Black currant are Consort, Crusader, Coronet, Ben Sarek, and Ben Nevis. Red currants and gooseberries are not affected by law and are legal to plant.
- There are no fungicides labeled on currants and gooseberries for control of white pine blister rust.

X-DISEASE

X-disease has been diagnosed in several peach orchards in northern region. Once the disease is established in an area or orchard, it can be very destructive. The disease has been reported to occur primarily in the Great Lakes states and in the province of Ontario, Canada. The range of distribution for X-disease corresponds to the range of distribution of the wild chokecherry, which is a major reservoir of infection. X-disease can affect peach, nectarine, sweet cherry, and sour cherry.

Symptoms

Symptoms on cherry and peach may be variable, depending upon the environment, variety, and possibly other factors. However, once acquainted with general symptoms, one should be able to recognize the disease more often than not.

Symptoms on Peach

This disease is most easily identified on peach. Symptoms

are predominantly foliar, but the fruits may also be affected. The first symptoms occur on foliage in midsummer.

Leaves on isolated branches curl inward and develop irregular, yellow to reddish or purple spots. The spots soon drop out, leaving a shothole effect and tattered leaves. Leaves on affected branches fall prematurely, starting at the base of the branch. Eventually, only a tuft of leaves remains at the tips of infected shoots.

A good diagnostic symptom is the presence of apparently healthy twigs or branches with normal-looking leaves mixed with twigs or branches showing the symptoms described earlier. This mixture of healthy and diseased branches on the same tree occurs primarily during the first and second years of infection. Two to three years after initial infection, most branches will show symptoms.

Another important diagnostic aid is to pull a tree and examine the roots. Foliar symptoms of yellowing and stunting can also be caused by Phytophthora root rot, which has also been observed in several northern orchards. With Phytophthora root rot, the above-ground symptoms are usually fairly uniform throughout the canopy of the tree. You generally will not see a healthy branch on the tree with all other branches showing symptoms.

In addition, the crown and/or roots of a tree with above-ground symptoms caused by root rot should have visible, rotted areas. These areas are generally brick-red to brown in colour and are often characterized by a sharp line of demarcation between healthy and diseased tissue. Roots on trees affected by X-disease appear normal.

Fruit set on trees infected with X-disease may appear normal at first, but fruit on infected branches will usually drop prematurely.

Symptoms on Cherry

There are two major types of reaction to X-disease in cherry trees. While there are exceptions in both cases, in general these two types can be described with reasonable safety as totally different reactions. The rootstock is the key.

Trees on mahaleb rootstock generally react differently to the disease than do those on mazzard rootstock. Cherries on mahaleb rootstock are killed suddenly in midsummer by the disease. Trees on mazzard rootstock decline slowly.

Infected sweet cherry trees on mazzard rootstocks may not show decline for many years; often the only recognizable symptom is on the fruit. Scattered fruit on trees propagated on mazzard rootstock are small and pink at harvest and have a bitter flavour. Sour cherries seem to be a little more seriously affected in that dieback and decline are often associated with the disease.

Cause

X-disease was believed to be caused by a virus until structures resembling mycoplasma were discovered in association with diseased plants. Mycoplasma are small parasitic organisms that have long been known to cause disease in plants, animals, and man. The organisms produce spherical- to ellipsoid-shaped bodies that are smaller than bacteria but larger than most virus particles. Mycoplasma live in phloem cells of plants. The phloem is a network of cells for moving foodstuffs manufactured in the leaves to other parts of the plant.

Spread of X-Disease

Fig. X-Disease

X-disease can be transmitted by several species of leafhoppers. These leafhoppers acquire the X-disease pathogen while sucking juices from the leaves of X-disease-infected choke-cherries.

Wild chokecherry is an important reservoir for the X-disease mycoplasma. Two to three weeks later the leafhopper can inject the agent into healthy leaves while feeding. These leafhoppers are usually not considered pests of peach and cherry. The main damage they cause is the incidental transmission of X-disease.

Movement of leafhoppers, and therefore X-disease, can also occur from infected sweet and sour cherry, particularly from trees on mazzard rootstock. Although the possibility of spread from peach to peach has been investigated many times, it appears to be of minor importance.

Control

- Control measures aimed at eradicating chokecherries near stone fruit orchards help to control X-disease. Chokecherry bushes are commonly found in hedges, along property lines, in open woods, and in overgrown meadows and abandoned fields.

Brush killers offer the cheapest and most effective way to kill the chokecherry bushes with both summer and autumn spray treatments. Other removal methods include bulldozing, deep plowing, burning, or pulling out individual bushes.

During the growing seasons following removal, check the treated area carefully for chokecherry sprouts. Sprouts or new chokecherry seedlings should be treated with herbicide sprays or pulled out during the summer.

- Reducing populations of leafhopper vectors is another approach to X-disease control. Maintaining a vigorous insect control programme from June through harvest with insecticides effective against leafhoppers should be useful in reducing spread of the disease.
- Remove infected cherry trees near peach orchards. X-disease is often very severe in young peach

orchards planted next to old cherry blocks with X-disease-infected trees on mazzard rootstock.

IDENTIFICATION OF CHOKECHERRIES

Chokecherry is often confused with wild black cherry and wild pin or fire cherry. It is important to be able to distinguish this species from others, because only chokecherry is important in the spread of X-disease. Unlike black cherry and pin cherry, which grow like trees to 50 feet or more, chokecherries are shrubs up to 15 feet tall. They are usually found in clumps. The fruit of chokecherry, which is produced along a central stem, is black when mature and ripens before those of black cherry.

The calyx cup on chokecherry fruit does not persist as it does on black cherry fruit. Pin cherry fruit are borne in clusters like sour cherry and in no way resemble those of chokecherry. Chokecherry leaves are wider and broader than black cherry or pin cherry leaves. Serrations along the margin of the leaves are more prominent and spreading than those of the other two species.

PHYTOPHTHORA ROOT ROT OF RASPBERRY

Phytophthora root rot is caused by several related species of soilborne fungi belonging to the genus Phytophthora. To date, P. megasperma, P. cryptogea, P. citriocola, P. cactorum, and at least two additional unidentified Phytophthora species have been implicated in this disease. The disease occurs on red, black, and purple raspberries, although in the Northeastern United States, it has been documented most commonly on red raspberries.

The disease has been reported to occur on blackberries in Kentucky. Phytophthora root rot can be an extremely destructive disease on susceptible cultivars where conditions favour its development. Infected plants become weak and stunted and are particularly susceptible to winter injury; seriously infected plants commonly collapse and die.

Symptoms

The disease is most commonly associated with heavy soils

or portions of the planting that are the slowest to drain (lower ends of rows, dips in the field, etc.). In fact, most declining plants that are considered to be suffering from "wet feet" may be suffering from Phytophthora root rot. Symptoms include a general lack of vigour and a sparse plant stand.

Apparently healthy canes may suddenly decline and collapse during the late spring or summer. In such cases, leaves may initially take on a yellow, red, or orange colour or may begin scorching along the edges. As the disease progresses, affected canes wilt and die. Infected plants frequently occur in patches, which may spread along the row if conditions remain favourable for disease development.

Because wilting and collapsing plants may be caused by other factors (winter injury, cane borers, etc.), it is necessary to examine the root system of infected plants to diagnose the disease. Suspect plants should be dug up and the epidermis (outer surface) scraped off the main roots and crown.

On healthy plants, the tissue just beneath the epidermis will be white; on plants with Phytophthora root rot, this tissue will be a characteristic brick red (eventually turning dark brown as the tissue decays). Sometimes a distinct line can be seen between infected and healthy tissue, especially on the below-ground portion of the crown.

In many fields, plants that are dying and declining because of Phytophthora root rot had previously been diagnosed as suffering from winter injury or "wet feet." One major difference in distinguishing between root rot and winter injury is that plants infected with Phytophthora root rot will continue to decline as time goes on and will not produce healthy primocanes, whereas winter-injured plants will usually send up healthy primocanes the year following the damaging winter.

DISEASE DEVELOPMENT

The fungi persist primarily as mycelium in infected roots or as dormant resting spores in the soil. When the soil is moist, reproductive structures (sporangia) are formed upon the infected tissue or by germinating resting spores (oospores) in the soil. Within each of these structures, a number of individual spores called zoospores are formed. These zoospores are expelled into the soil during periods when the soil is saturated with water.

The zoospores have "tails" (flagella) that allow them to swim through the water-filled soil pores to reach new plant parts. Upon reaching a plant root or crown, the zoospores become attached and begin the infective process. As water remains standing and oxygen is depleted from the root zone, the plant is progressively less capable of resisting the fungus's attempts at invasion, and infection becomes more likely and severe.

Each new infection site is a potential source of additional resting spores and zoospores, allowing for epidemic disease development in sites that are subjected to repeated periods of standing water. Although the optimum season for infection is not known for certain, it is likely that spring and fall are particularly favourable periods. However, it is assumed that infection can occur throughout the growing season if soil moisture conditions are favourable.

Control

There is no one simple "cure" for this disease. However, there are a number of different practices or methods that growers can use to avoid or minimize losses. Because no single method is completely effective by itself, the best strategy is to develop an integrated disease management programme, where as many control practices as possible are used within an integrated approach. These methods should be considered and used:

Exclusion

Avoid introducing the Phytophthora fungi if you are planting into an uninfested site, especially one that has not previously contained fruit crops. Circumstantial evidence suggests that symptomless nursery plants may be an important means of initially introducing this pathogen into a grower's field.

Fortunately, it is now possible to buy many different raspberry cultivars that have been propagated through tissue culture techniques in a laboratory and greenhouse, without ever coming into contact with field soil. Such plants pose little risk of introducing Phytophthora fungi into the field. To minimize your risk of setting out contaminated plants, use only those that come directly from the laboratory or greenhouse and have not been grown out in nursery fields before sale.

Drainage

Any practice that will prevent water from collecting around plants will reduce the incidence and severity of Phytophthora root rot. This includes both good planting-site selection and site modification when necessary. Included in site modification are the placement of tile drains and growing plants on raised beds. Using a raised-bed planting system can provide substantial control of Phytophthora root rot of raspberry.

Resistance

One of the best techniques for controlling any disease is the planting of resistant varieties and the avoidance of highly susceptible varieties. Phytophthora root rot is most serious on red raspberries and some of the hybrids. The black raspberry varieties 'Cumberland' and 'Munger' are reported to be susceptible. The cultivars 'Bristol,' 'Dundee,' and 'Jewel' appear to be moderately to highly resistant.

Among red raspberry cultivars, none are immune to the disease, but cultivars do differ greatly in their level of susceptibility. Among varieties grown in the Midwest and Northeast, 'Titan' and 'Hilton' are extremely susceptible, with 'Festival,' 'Heritage,' 'Reveille,' and 'Taylor' moderately to highly susceptible. 'Newburgh' is somewhat resistant, and 'Latham,' 'Boyne,' 'Killarney,' and 'Nordic' are considered to be fairly resistant.

Fungicide

Phytophthora root rot of raspberry can be partially

controlled with the soil-applied fungicide Ridomil Gold. Although Ridomil is effective for control of Phytophthora root rot, it should be remembered that it is merely an additional disease-management tool. It will give you the best results only when the other disease management practices that have been discussed are also followed.

ANTHRACNOSE OF GRAPE

Anthracnose of grape was first detected in the United States in the mid 1800s. The disease was probably introduced into this country by grape plant material imported from Europe. It quickly established in American vineyards and became a significant disease of grape in rainy, humid, and warm regions of the United States.

The disease is not common; however, it caused severe damage in a central vineyard on the cultivar 'Vidal' in 1993 and in a southern vineyard on the cultivars 'Vidal' and 'Reliance' in 1998. Anthracnose reduces the quality and quantity of fruit and weakens the vine. Once the disease is established in a vineyard, it can be very destructive.

Symptoms

All succulent parts of the plant, including fruit stems, leaves, petioles, tendrils, young shoots, and berries, can be attacked, but lesions on shoots and berries are most common and distinctive. Symptoms on young, succulent shoots first appear as numerous small, circular, and reddish spots. Spots then enlarge, become sunken, and produce lesions with gray centres and round or angular edges.

Dark reddish-brown to violet-black margins eventually surround the lesions. Lesions may coalesce, causing a blighting or killing of the shoot. A slightly raised area may form around the edge of the lesion. Infected areas may crack, causing shoots to become brittle.

Anthracnose lesions on shoots may be confused with hail injury; however, unlike hail damage, the edges of the wounds caused by the anthracnose fungus are raised and black. In addition, hail damage generally appears on only one side of the shoot, whereas anthracnose is more generally distributed. Anthracnose on petioles appears similar to that on the shoots.

Leaf spots are often numerous and develop in a similar manner to those on shoots. Eventually, they become circular with gray centres and brown to black margins with round or angular edges. The necrotic centre of the lesion often drops out, creating a shot-hole appearance.

Young leaves are more susceptible to infection than older leaves. When veins are affected, especially on young leaves, the lesions prevent normal development, resulting in malformation or complete drying or burning of the leaf. Lesions may cover the entire leaf blade or appear mainly along the veins.

On berries, small, reddish circular spots initially develop. The spots then enlarge to an average diameter of 1/4 inch and may become slightly sunken. The centres of the spots turn whitish gray and are surrounded by narrow reddish-brown to black margins. This typical symptom on fruit often resembles a bird's eye, and the disease has been called bird's eye rot. Acervuli (fungal fruiting structures) eventually develop in the lesions. A pinkish mass of fungal spores (conidia) exudes from these structures during prolonged wet weather.

Causal Organism

Anthracnose of grape is caused by the fungus *Elsinoe ampelina*. The fungus overwinters in the vineyards as sclerotia (fungal survival structures) on infected shoots. In the spring, sclerotia on infected shoots germinate to produce abundant spores (conidia) when they are wet for 24 hours or more and the temperature is above 36°F. Conidia are spread by splashing rain to new growing tissues and are not carried by wind alone.

Another type of spore, called an ascospore, is produced within sexual fruiting bodies and may also form on infected canes and berries left on the ground or in the trellis from the previous year. The importance of ascospores in disease development is not clearly understood.

Conidia are by far the most important source of primary

inoculum in the spring. In early spring, when free moisture from rain or dew is present, conidia germinate and infect succulent tissue. Conidia germinate and infect at temperatures ranging from 36 to 90°F.

The higher the temperature, the faster disease develops. Disease symptoms start to develop approximately 13 days after infection occurs at 36ºF and at four days after infection occurs at 90°F. Heavy rainfall and warm temperatures are ideal for disease development and spread.

Lesions may extend into the pulp and cause the fruit to crack. Lesions on the rachis and pedicels appear similar to those on shoots. Clusters are susceptible to infection before flowering and until *Véraison*.

Once the disease is established, asexual fruiting bodies called acervuli form on diseased areas. These acervuli produce conidia during periods of wet weather. These conidia are the secondary source of inoculum and are responsible for continued spread of the fungus and the disease throughout the growing season.

Disease Management

- Sanitation is very important. Prune out and destroy (remove from the vineyard) diseased plant parts during the dormant season. This includes infected shoots, cluster stems, and berries. This should reduce the amount of primary inoculum for the disease in the vineyard.
- Eliminate wild grapes near the vineyard. The disease can infect wild grapes, and infected wild grapes have been observed near diseased vineyards. Wild grapes provide an excellent place for the disease to develop and serve as a reservoir for the disease. It is probably impossible to eradicate wild grapes from the woods, but serious efforts should be made to at least remove them from the fence rows and as far away from the vineyard as possible.

  Remember, the spores are spread over relatively short distances by splashing rain and should not be able to move over long distances by wind into the vineyard.

- Varieties differ in their susceptibility. The disease has been observed on 'Vidal' and 'Reliance.' Vinifera and French Hybrid cultivars may be more susceptible than American grapes, such as 'Concord' and 'Niagara.'
- Canopy management can aid in disease control. Any practice that opens the canopy to improve air circulation and reduce drying time of susceptible tissue is beneficial for disease control. These practices include selection of the proper training system, shoot positioning, and leaf removal.
- Fungicide use. Where the disease is established, especially in a commercial vineyard, the use of fungicides is recommended. Fungicide recommendations for anthracnose control consist of a dormant application of Liquid Lime Sulfur in early spring, followed by applications of foliar fungicides during the growing season.

ANTHRACNOSE OF STRAWBERRY

Anthracnose is an important disease of strawberry that can affect foliage, runners, crowns, and fruit. The disease is caused by several species of fungi in the genus Colletotrichum: Colletotrichum acutatum, Colletotrichum fragariae, and Colletotrichum gloesporoides. They all cause similar or nearly identical symptoms on strawberry. The two most destructive forms of the disease are crown rot, usually associated with Colletotrichum fragariae, and fruit rot, usually associated with Colletotrichum acutatum.

Historically, anthracnose has generally been restricted to the southern United States and was not common in the northern United States. It has generally been considered to be a "warm weather" or "southern disease" of strawberry. Epidemics of anthracnose fruit rot caused by Colletotrichum acutatum have occurred, but the crown rot phase has been observed only a few times in the mid 1980s.

Over the past few years, the incidence of anthracnose fruit rot in northern production areas has increased, and there is a concern about the potential impact of this disease in northern, perennial-production systems. Although the disease occurs sporadically and is not common in most plantings, when it does occur, it can be devastating, resulting in 100 per cent loss of fruit.

Symptoms

The fungus can attack fruit, runners, petioles, and the crown of the plant. Dark elongated lesions develop on petioles and runner stems. Affected petioles and stems are sometimes girdled by lesions, causing individual leaves or entire daughter plants to wilt and die.

Under warm, humid conditions, salmon-coloured masses of spores may form on the lesion surface. If the crown tissue is infected, crown rot may develop and the entire plant may wilt and die. When infected crowns are sectioned, internal tissue is firm and reddish-brown to dark-brown in colour. Crown tissue may be uniformly discoloured or streaked with brown.

Whitish, tan, or light-brown water-soaked lesions up to 3 mm in diameter initially develop on fruit. The lesions eventually turn brown or dark-brown, are sunken, and enlarge within two to three days to cover most of the fruit. Lesions are covered with pale-orange or salmon-coloured spore masses. Under moist conditions, the fungus may grow out around the edge of the lesion or through the lesion, giving a fuzzy appearance. Infected fruit eventually dry down to form hard, black, shriveled mummies. Fruit can be infected at any stage of development.

Disease Development

Although anthracnose can be caused by several species of fungi in the genus *Colletotrichum, Colletotrichum acutatum* is the most common species causing fruit rot. The disease is probably introduced into new plantings on infected plants. Recent research indicates that the fungus can grow and produce spores on the surface of apparently healthy leaves. Once the disease is established in the field, the fungus can overwinter on infected plants and plant debris, such as old dead leaves and mummified fruit.

Spore production, spore germination, and infection of strawberry fruits are favoured by warm, humid weather and rainfall. In spring and early summer, spores are produced in abundance on previously infected plant debris. The spores are spread by splashing rain, wind-driven rain, and by people or equipment moving through the field. They are not airborne so they do not spread over long distances in the wind. Spores require free water on the plant surface in order to germinate and infect.

The optimum temperature for infection on both immature and mature fruit is between 77 and 86°F. Under favourable conditions, the fungus produces secondary spores on infected fruit. These spores are spread by rain and result in new infections throughout the growing season. Disease development can occur very rapidly. Up to 90 per cent of the fruit can be infected within a week or less. Both immature and mature fruit are susceptible to infection; however, the disease is most common on ripening or mature fruit.

Disease Management

- Use disease-free planting material. The disease is introduced to the field with infected plant material. The best way to avoid the disease is to begin with disease-free planting material. Although there are no nurseries that can certify plants to be free of fungal and bacterial plant pathogens, inspection of plants for the disease before planting is recommended.
- Proper irrigation. If the field was previously infected, or the disease is present in the field, minimize the amount of overhead irrigation used. The fungus is spread by splashing water. Avoid the use of overhead irrigation and use drip irrigation if possible.
- Mulching. Plastic mulch increases the level of splash dispersal of the pathogen. Mulching with straw is

recommended in perennial matted row plantings to reduce water splash and disease spread.

- Remove infected plant parts. Infected plant parts serve as a source of inoculum for the disease. Remove as much old, infected plant debris as possible. Try to remove infected berries from the planting during harvest.
- Fungicide use. Once anthracnose fruit rot is established in a planting, it is difficult to control with fungicides. Fungicides for control of anthracnose fruit rot should be used in a protectant or preventative programme. In order to obtain effective disease control, fungicides should be applied before the disease develops.

LATE LEAF RUST OF RED RASPBERRIES

Late leaf rust is a potentially serious disease of red raspberries. Late leaf rust does not affect black raspberries or blackberries. The disease can affect leaves, canes, petioles, and fruit. Economic losses occur from fruit infection and premature defoliation. Because it usually appears late in the season, and only occasionally in a severe form, some consider it to be a minor disease.

However, losses due to fruit infection have reached 30 per cent in commercial red raspberry plantings. The wild red raspberry, *Rubus strigosus*, in the eastern United States is very susceptible to this disease. A number of cultivated varieties originating from this species also are highly susceptible. While late leaf rust occurs throughout the northern half of the United States and southern Canada, it is more common east of the Mississippi River. In recent years, its occurrence has increased in the northern areas of the Midwest, and it has caused significant losses.

On mature leaves, small chlorotic, or yellow, areas initially form on the upper surface of infected leaves. These spots may eventually turn brown before leaves die in the fall. Unless the disease is severe, foliar infections may be difficult to see. Small pustules filled with yellow to orange powdery spores (not waxy like the spores of orange rust) are formed on the underside of infected leaves. Badly infected leaves may drop prematurely, and in years when the disease is severe, canes may be bare by September.

Flower calyces, petioles, and fruit at all stages of development may be attacked. On fruit, pustules develop on individual drupelets, producing yellow masses of spores, which make the berries unattractive and unacceptable for fresh market sales. Infections may also occur on leaf petioles and canes.

Late leaf rust is caused by the fungus *Pucciniastrum americanum*. Unlike the fungus that causes orange rust, the late leaf rust fungus is not systemic. The fungus is heteroecious, meaning that it attacks two different hosts at different stages of its life cycle. The rust fungus produces two types of spores only on red raspberries. The alternate host for the rust is white spruce (*Picea americanum*), on which another type of spore (aeciospore) is produced.

Aeciospores are released from infected white spruce in mid-June to early July and are capable of infecting raspberry during this period. In early July, urediniospores (powdery-yellow to orange spores) start to form on the underside of infected raspberry leaves or flower parts. These urediniospores can continue to cause infections on raspberry leaves and fruit throughout the growing season.

Another type of spore (teliospore) develops on infected leaves in the fall and serves as the overwintering form of the fungus. In the following year, the teliospores germinate and form yet another type of spore (basidiospore), which infects white spruce needles during rainy periods from mid-May to early June.

Several recent studies indicate that the fungus apparently does not need the aeciospore stage to survive on raspberries, because the disease is found year after year in regions remote from any spruce trees. It is probable that the fungus overwinters on infected raspberry canes as urediniospores or teliospores that serve as the source of primary inoculum for new infections the following season.

- Use healthy, disease-free planting stock. One of the best ways to avoid the disease is to start the planting with healthy planting stock. Since the fungus can be carried in or on planting material, inspection of the planting materials before planting is recommended.
- Site selection. Select a site with good air movement and full sun exposure. Never plant raspberries in shaded areas. Good air movement and sunlight help the foliage and fruit to dry off quickly after a rain or heavy dew. Rapid drying will reduce the incidence of fruit and leaf diseases in general.
- Canopy management. Keep row width between 1 and 2 feet in order to encourage air movement and faster drying. Cane density should not exceed three or four canes per square foot. Always select large, healthy canes when thinning. Control timing and the amount of nitrogen fertilizer to prevent excessive growth.
- Control weeds. Good weed control within and between the rows is essential. Weeds in the planting prevent air circulation and increase drying time, resulting in wet fruit and foliage for longer periods.
- Sanitation. Remove and destroy infected and old fruited canes. Previously infected plant parts serve as a source of inoculum for the disease. Removing and destroying old fruited and infected primo canes greatly reduces the amount of disease inoculum in the planting.
- Eradication of alternative and wild hosts. As previously mentioned, the late leaf rust fungus requires white spruce trees as an alternate host to complete its full life cycle. Eradication of white spruce trees interrupts the life cycle of the fungus and should aid in disease control. Eradication of nearby wild red raspberries that serve as a reservoir for disease is also beneficial for control of the disease.
- Use of disease resistance. Black raspberries and blackberries are immune to the disease.

Unfortunately, there are no commonly grown red raspberry varieties that are resistant to the disease.

- Fungicide use. Fungicides that are effective for control of late leaf rust are currently available and are commonly used in commercial plantings.

PHOMOPSIS LEAF BLIGHT AND FRUIT ROT OF STRAWBERRY

Phomopsis leaf blight is a common disease of straw-berry in the eastern United States Although the fungus infects leaves early in the growing season, leaf blight symptoms are most apparent on older leaves near or after harvest.

The economic importance of leaf blight appears to be relatively minor; however, incidence of the disease has been increasing. The disease can weaken strawberry plants through the destruction of older foliage. Weakened plants can result in reduced yields the following year. In years highly favourable for disease development, leaf blight can cause defoliation and, in some cases, death of plants.

Especially in warmer climates, the fungus that causes leaf blight can also cause a fruit rot called soft rot. The first observation of Phomopsis fruit rot (soft rot) was on plants growing under plastic culture in 1999. Although not common. Phomopsis fruit rot can result in serious losses.

Symptoms

Leaf lesions start as circular spots that are similar to that of leaf spot disease of strawberry; however, young lesions caused by Phomopsis often have a reddish halo that is not present on leaf spot lesions. As the disease progresses, irregular, often circular, zoned lesions may form. Typically, there are three zones in a lesion: a dark brown centre, surrounded by a tan to light brown zone, which is surrounded by a reddish or purplish outer zone. In later stages of disease development, lesions, especially ones along veins, may become V-shaped, with the widest part of the V toward the margin of the leaflet. These V-shaped lesions are characteristic of the disease.

Symptoms develop on ripening or fully matured fruits. Early symptoms are round, pink, water-soaked spots. Eventually, spots enlarge and turn brown (dark brown in the centre with a lighter brown edge). The texture of lesions also changes from water-soaked to a brown "crusty" surface.

The brown crusty areas are actually clusters of tiny fruiting structures (pycnidia) produced by the fungus. These pycnidia can be observed with the aid of a 10x hand lens or magnifying glass. In later stages of development, fruit rot symptoms resemble that of anthracnose fruit rot. Lesions of anthracnose fruit rot do not develop pycnidia.

Causal Organism and Disease Cycle

Phomopsis leaf blight and fruit rot (soft rot) of strawberry are caused by the fungus, *Phomopsis obscurans*. The fungus survives overwinter within infested plant debris or within infected plant parts. The fungus produces pycnidia on old infected tissues.

Each pycnidia contains thousands of spores (conidia) that ooze out under moist conditions and are disseminated to other plant parts by splashing water (rain or irrigation). In early spring, these conidia infect leaves or other plant parts and cause new infections. Free water on the plant surface is required for the conidia to germinate and infect the plants.

Although leaves are infected early in the growing season, symptoms generally do not develop until later in the season. The fungus remains symptomless in leaves as a "latent" or dormant infection. Later in the growing season, often during or after harvest, the fungus becomes active and leaf lesions develop.

If lesions develop prior to harvest, they can produce pycnidia which release conidia to infect the fruit. Disease development is favoured by long wet periods (15 hours or more). Temperature has less effect on disease development than wetness duration. In fact, the fungus can cause infection over a wide range of temperatures.

Selection of planting material: Always use certified disease-free plants. Varieties appear to differ in their susceptibility to leaf blight; however, specific information on disease resistance is generally not available. Varieties that appear to be highly susceptible to leaf blight at your location should be avoided.

Sanitation: Removal of old infected leaves and other plant parts in which the fungus overwinters will decrease the amount of fungal inoculum available to initiate the disease the following spring.

Site selection: Always plant strawberries on a site with excellent soil drainage and air circulation. Since the fungus requires long wetness periods to infect plants, any practice that promotes faster drying of plant parts is beneficial to disease control. Always plant in a location with full (all-day) sun. Never plant in shaded areas. Control weeds: Controlling weeds within the planting is an important cultural practice for successful strawberry production. In addition, weeds prevent air circulation in the planting, resulting in plants staying wet for longer periods of time.

Fungicide use: Fungicides are currently available that have good activity for control of leaf blight; however, emphasis for control should be placed on the use of cultural practices and avoiding highly susceptible varieties.

WOOD ROT

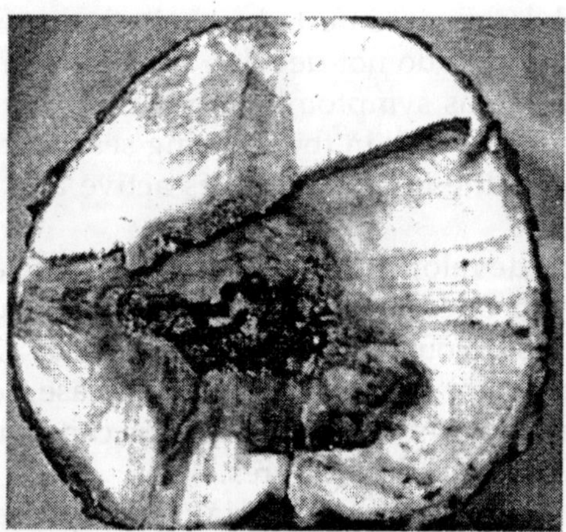

Fig. Wood Rot

Many homeowners are familiar with wood damage caused by rot. They see it in structural lumber, log homes, eaves, garage doors, exterior door trim, window casings and other wood used in construction. Current estimates show that replacement materials, needed to repair damage caused by rot alone, account for nearly 10 per cent of U.S. annual wood production.

Blame for this destruction is sometimes incorrectly placed on termites or other wood-destroying insects. However, there are no mud tunnels or mines in the wood such as seen with termite and other wood-destroying insects, nor is there any sawdust, which would be evident in the case of carpenter ant damage.

The key to preventing rot is to control the wood's exposure to moisture and to employ an effective prevention and treatment programme. Most wood decay fungi grow only on wood with a high moisture content, usually 20 per cent or above. Green (unseasoned) lumber is a prime target for decay fungi.

There are two main classes of wood rot. In one type, the decayed area has a brown discolouration and a crumbly appearance. It usually breaks up into variously-sized cubes, giving rise to the name "brown cubical rot." Another type of rot results in a white or yellow discolouration, with the decayed wood being "stringy" or "spongy."

Although many decay fungi may grow for long periods without producing any external evidence of their presence, others produce "fruiting bodies" on the surface of decaying wood. Fruiting bodies are usually "crusts" or shelflike "brackets" which are a few inches or so in diameter. The fruiting body of *Serpula lacrimans,* e.g., is a rust-brown, crust-like structure on the wood surface.

It has a waxy appearance, with shallow, net-like folds or "wrinkles." The fruiting body of *Poria incrassata* is also crust-like. It is white to light buff when initially formed, but becomes brown as it ages and dries out. Small pores can be seen in the crust when it is examined with a hand lens. *Gleophyllum trabeum* forms bracket-like fruiting bodies.

The upper surface of the fruiting body is dull gray-brown and smooth. The lower surface has elongate openings (pores) or split-like openings (gills). These fruiting bodies produce millions of tiny spores which may, in some cases, serve to spread the decay fungus to other areas.

Also, surface molds, "mildews," and stain fungi are often found growing on the surface of damp wood and can be confused with decay fungi. Although these organisms may discolour the wood, they do not break down wood fibres and thus do not weaken its structure.

However, these organisms indicate that moisture is present and that decay will likely proceed if a wood-rotting fungus becomes established in the wood.

Chapter 8

Control of Plant Disease

Decay fungi are living organisms which send minute threads called "hyphae" through damp wood, taking their food from the wood as they grow. Gradually, the wood is decomposed and its strength is lost. Such damage is often inconspicuous until its final stages, and in a few instances homeowners have suddenly found floors breaking through or doors falling from their hinges due to wood rot. When previously dry wood is placed in contact with moist soil, or in a location where it is subject to condensation (such as unventilated crawl space), it is likely that wood decay problems will occur.

Rain leaks, faulty plumbing and leaky downspouts also are common sources of moisture. In some instances, water can be transported to the site of decay through strands or "rhizomorphs" of the decay fungi. Water-transporting strands may extend for thirty or more feet across brick, concrete or similar materials.

The wood decay fungus, *Serpula lacrimans*, has been known to transport water up three stories to an area where decay is occurring. *Poria incrassata* is also capable of transporting water long distances. However, these fungi are exceptions to the rule. Most wood-rotting fungi must have a direct supply of water at the site of decay. Thus the term "dry-rot," sometimes applied to decay in wood structures, is erroneous.

Prevention

- If the decay hazard is high, select the heartwood of

decay-resistant species or use wood properly treated with a good preservative. Conifers from which decay-resistant lumber is produced include Pacific yew, juniper, redwood, baldcypress, and western red cedar. Durable hardwood species include osage orange, black locust, red mulberry, catalpa and black walnut.

- Build on a well-drained site. Use proper grading to prevent water from seeping under the house. Install effective drain tile, roof overhang, gutters, and downspouts. Place no untreated wood within 18 inches of the ground.
- Provide adequate cross ventilation beneath buildings to eliminate dead air pockets. Install two square feet of opening for 25 linear feet of wall. Dense bushes or other plants should not be placed in front of these ventilators.
- Install a vapour barrier on the soil surface to cause soil moisture to condense on the barrier and return to the soil rather than condensing on the floor and above joists. Satisfactory barriers can be made by covering the soil with asphalt roofing paper or polyethylene sheets.

Repair of Decayed Buildings

First determine the source of moisture and remove it. If adequate ventilation and soil drainage are provided and all contacts of untreated wood with the soil or moist concrete or masonry are broken, decayed wood will dry out and further decay will be stopped. When making replacements, cut out at least one foot beyond the rotten area. Avoid placing new lumber in contact with old, decayed wood. Replacement lumber should be treated before installation. Remodel to provide more ventilation and better design rather than simply replacing decayed lumber.

BACTERIAL CROWN GALL OF FRUIT CROPS

Crown gall is caused by the bacterium *Agrobacterium*

tumefaciens. This bacterium has the widest host range of any plant pathogen. It is capable of causing tumors, or "galls," on virtually all plant species, except the monocots (grasses). A similar bacterium, *Agrobacterium rubi*, causes galls on the canes of brambles. All fruit crops grown are susceptible.

The disease is particularly destructive on brambles (raspberries and blackberries) and grapes. It can also cause severe problems on apple, pear, blueberry, all stone fruits and on ornamentals. The bacteria induce galls or tumors on the roots, crowns, trunks and canes of infected plants. These galls interfere with water and nutrient flow in the plants. Seriously infected plants may become weakened, stunted and unproductive.

Symptoms

The disease first appears as small overgrowths or galls on the roots, crown, trunk or canes. Galls usually develop on the crown or trunk of the plant near the soil line or underground on the roots. Above ground or aerial galls may form on canes of brambles and highly susceptible cultivars of grape. Although they can occur, aerial galls are not common on fruit trees. In early stages of development the galls appear as tumor-like swellings that are more or less spherical, white or flesh-coloured, rough, spongy (soft) and wart-like. They usually form in late spring or early summer and can be formed each season.

As galls age they become dark brown to black, hard, rough, and woody. Some disintegrate with time and others may remain for the life of the plant. The tops of infected plants may appear normal. If infection is severe, plants may be stunted, produce dry, poorly-developed fruit, or show various deficiency symptoms due to impaired uptake and transport of nutrients and water.

Causal Organism

The crown gall bacterium is soil-borne and persists for long periods of time in the soil in plant debris. It requires a fresh wound in order to infect and initiate gall formation.

Wounds that commonly serve as infection sites are those made during pruning, machinery operations, freezing injury, growth cracks, soil insects and any other factor that causes injury to plant tissues.

Bacteria are abundant in the outer portions of primary galls, which is often sloughed off into the soil. In addition to primary galls, secondary galls may also form around other wounds and on other portions of the plant in the absence of the bacterium. The bacteria overwinter inside the plant (systemically) in galls, or in the soil. When they come in contact with wounded tissue of a susceptible host, they enter the plant and induce gall formation, thus completing the disease cycle. The bacteria are most commonly introduced into a planting site on or in planting material.

Control

- Obtain clean (disease free) nursery stock from a reputable nursery and inspect the roots and crowns yourself to make sure they are free from galls. Avoid planting clean material in sites previously infested with the bacteria.
- Avoid all unnecessary root, crown and trunk wounding by careless cultivation and other machinery operation, and control soil insects. Any practice that reduces wounding is highly beneficial. Preventing winter injury (especially on grapes) is also beneficial.
- On grapes, the double trunk system of training may be a useful system for minimizing losses due to crown gall. If one trunk is infected, it can be removed. The remaining trunk can be pruned leaving a full number of buds until the second trunk can be renewed. Galls on the upper parts of the trunk or on canes can be removed by pruning.
- A relatively new biological control agent for crown gall is available for apple, pear, stone fruit, blueberry, brambles and many ornamentals. It is not effective on grape. The agent is a nonpathogenic strain of

bacterium (*Agrobacterium radiobacter* strain 84) that protects the plants against infection by the naturally occurring strains of pathogenic bacteria in the soil. Nursery stock is dipped in a suspension of commercially prepared *Agrobacterium radiobacter* strain 84 at planting time. The antagonistic bacteria act only to protect disease free plants from future infection by the crown gall bacterium; they cannot cure infected plants.

BITTER ROT OF APPLES

Bitter rot is a common disease of apples and pears in practically all countries where they are commercially grown. Of the three fruit rot diseases on apple (bitter rot, white rot and black rot), bitter rot has the potential to be the most destructive. The fungus that causes bitter rot (fruit rot) can also cause a leaf spot and canker, although the leaf spot and canker form of the disease are not common. The disease is most common in warmer regions because high temperatures favour disease development. Before the development of effective fungicides, entire crops were lost to bitter rot during periods of warm, wet weather.

Symptoms

Fruit rot symptoms differ somewhat, depending on whether infection is initiated by spores from perithecial strains of the fungus (strains that produce ascospores and conidia) or conidial strains of the fungus (strains that produce only conidia). Initial symptoms produced by perithecial or conidial strains are similar. Lesions begin as small, slightly sunken areas, which are light brown to dark brown.

On mature fruit, lesions may be surrounded by a red halo. Lesions originating from infections by conidial strains remain circular and become sunken as they enlarge. Copious quantities of ooze containing conidia (spores) are produced in fungal fruiting bodies called acervuli, which occur in concentric circles on rotted tissue around the point of infection.

Acervuli are sparse on some lesions and very dense on

others. Under moist, humid conditions, the spore masses appear creamy and are salmon to pink in colour. Lesions initiated by perithecial strains are usually not sunken and are often darker brown than those caused by conidial strains. Acervuli are widely scattered over the surface, and perithecia are found in dark brown to black clumps scattered on the surface.

Bitter rot lesions on all fruit (regardless of which strain caused infection) extend in a cone shape manner toward the core. In cross section, the lesion appears "t-shaped". This is a reasonably reliable characteristic that can be used to separate bitter rot from white rot or black rot.

The rotten area is brown but much firmer than white rot. Infected fruit eventually mummify, and some may remain attached to the tree through the winter. Leaf lesions are not common and are caused by the perithecial strain. They begin as small, red flecks, which enlarge to irregular brown spots 1/16 to 1/2 inch in diameter. Severely affected leaves may drop off the tree. Bitter rot cankers are rare in the eastern United States. Cankers are oval, sunken and often zonate in appearance.

Casual Organism and Disease Cycle

Bitter rot is caused by the fungus, *Glomerella cingulata*. The fungus overwinters in apple orchards in dead wood or mummified fruit in the trees that were infected during the previous season. Conidia, produced in these overwintering sites, are the primary inoculum source in the spring, although ascospore inoculum is important in some orchards. Conidia are spread by splashing and wind-blown rain. Insects and birds are also involved in their dispersal.

Ascospores are released after rain and are airborne. Fruit are susceptible to infection from 3 weeks after petal fall until harvest. Temperatures of 80 or 90 degrees F are most favourable for disease development. Because of the large number of conidia produced in lesions on fruit, the fungus has the potential for rapid spread within the orchard.

Control

- Control of bitter rot is best achieved through an

integrated programme of cultural practices and chemical control measures.

- Sanitation is critical for effective control. Piles of prunings are an important source of inoculum and should be removed from the perimeter of the orchard or burned. Prunings can be left on the orchard floor if they are chopped with a flail mower, which removes much of the bark and allows them to decompose faster.

  Removal of mummified apples and pruning out dead wood in the tree are important for reducing the inoculum within the tree. Pruning out current-season shoots infected with fire blight is also important, because they can be colonized and serve as an inoculum source during the same growing season.
- Any practice that helps to maintain trees in a healthy vigorous condition is critical for controlling the canker phase of the disease. Cankers generally develop only on stressed or weakened trees. Prune trees annually and maintain a balanced fertility programme based on soil and foliar nutrient analysis. Cankers generally develop rapidly on winter-injured trees.
- The use of fungicides combined with good sanitation is beneficial for controlling the fruit rot phase of the disease. Fungicides are not effective for controlling the canker phase of the disease on weakened trees.

Wild Mushrooms

There are 2,000 or more kinds of wild mushrooms. Some are poisonous and some are edible and delicious when properly prepared. The edibility of the majority is either not known or they are not considered for food because of their small size or poor flavour or texture.

Even though not every one is interested in collecting mushrooms to eat, it is important to understand most have an important and beneficial role in the environment. They grow in a wide variety of habitats. Most of the mushrooms seen on a walk through a woods are beneficial. Many species are quite specific about their food source and will be found only under or near certain kinds of trees-some under pines, others under oak, etc. Some are important as decay organisms, aiding in the breakdown of logs, leaves, stems and other organic debris.

Fig. Wild Mushrooms

This important role of mushrooms results in recycling of essential nutrients. Some mushrooms grow in, and form their fruiting structures on living trees causing decay of the sapwood or of the heartwood. Many woodland mushrooms are essential to good growth, and even survival of trees. They establish a relationship with roots of living trees that is mutually beneficial. These are called mycorrhizal mushrooms.

All mushrooms, whether poisonous or edible can be admired for their beauty and the fantastic variety of form, colour and texture.

WHICH MUSHROOMS ARE SAFE TO EAT?

Some edible mushrooms are very similar in appearance to poisonous kinds and may grow in the same habitat. Edible mushrooms are known to be safe to eat because they have been eaten frequently with no ill effects. Poisonous mushrooms are known because someone ate them and became ill or died. There is no test or characteristic to distinguish edible from poisonous mushrooms. This indicates a need to identify with certainty one of several of the proven edible species and pick and eat only those positively identified.

At the same time, you should also learn to identify some of the common poisonous mushrooms, especially those that are similar to edible kinds. It is especially important to learn the characteristics of the Amanita mushrooms, since several of the species common are poisonous, a few causing serious illness and sometimes death.

The word "toadstool" is often used to indicate a poisonous mushroom. Since there is no way to distinguish between a so-called "toadstool" and an edible mushroom it is more precise to speak of poisonous mushrooms or edible mushrooms. The season for collecting wild mushrooms for food begins in late March and early April when the first morel or sponge mushrooms are found.

These choice edible mushrooms are most abundant during April and the first two weeks of May. The false morels (members of the Gyromitra genus) are found at this same time of the year, but they must be regarded as poisonous and not collected for eating. It is true that many have eaten false morels with no apparent ill effects. However, recent research has shown toxins to be present in some of the false morels that can cause death or serious illness. Do not eat the false morels.

From mid summer to late autumn, a great variety of mushrooms may be found. A number of these are choice edibles. Photographs and brief descriptions of several of the more common mushrooms found are included in this fact sheet.

Control of Nuisance and Detrimental Molds (Fungi) in Mulches and Composts

Mulches and composts are often used to improve soils and plant health and to control weeds. They improve drainage as they decompose even though the ability of the soil to hold moisture is increased. They lower soil temperature in the summer and insulate roots from cold in winter conditions. Eventually, they mineralize, release nutrients for plants, and leave humic substances as residues. Their beneficial side effects gradually disappear unless more mulch or compost is applied.

Generally, these organic materials inhibit undesirable microorganisms such as soilborne pathogens that cause diseases of plants. They also stimulate the activity of many types of beneficial microorganisms, including mycorrhizal fungi. Occasionally, however, microorganisms (primarily fungi) in mulches and composts can become a nuisance and even cause certain diseases of plants.

Whether a mulch or a compost provides beneficial or detrimental effects is largely determined by the type of organic matter from which it was produced and the degree to which it was decomposed and treated before its application in the landscape. The temperature, pH, and moisture content of the products just before application also have an effect. The severity of nuisance fungi can be minimized if appropriate steps are taken in time.

Examples of Nuisance Fungi

The shotgun or artillery fungus (*Sphaerobolus*) may cause serious problems. While it decays the mulch, it also produces fruiting structures that resemble tiny cream or orange-brown cups that hold a spore mass resembling a tiny black egg (1/10 inch in diameter). This fungus shoots these spore masses high into the air. They stick to any surface and resemble small tar spots on leaves of plants or the siding of homes. They are difficult to remove, leave stained surfaces, and may result in major damage.

Slime molds are another type of nuisance fungus. They first appear as bright yellow or orange slimy masses that may be several inches to a foot or more across. They produce tiny spores that eventually dry out and blow away. These molds, like many others such as stink horns and bird's nest fungi, actually should be considered microbial ornamentals in the landscape. However, some fungi in mulches and composts produce toad stools (mushrooms), and some of these are toxic to humans. It is a good idea to destroy them when small children have access to the mulched area.

Another fungal problem that is often not identified correctly occurs when mulches are applied too deep (4-6 inches) instead of the ideal depth of 1.5 to 2 inches. Deep layers of mulch, particularly if prepared from fresh woody materials, may actually undergo high temperature decomposition during the summer.

The result is that the mulch dries out to less than 34 per cent moisture and becomes a dusty mass. Fungi often colonize these dry mulches until they become a water-repelling, moldy chunk of material. Young trees mulched in this way sometimes die from drought even though the homeowner irrigates the area, because water runs off the mulch since it repels water.

Other Types of Problems

Fresh mulches prepared from trees killed by plant diseases may be colonized by plant pathogens. *Verticillium dahliae,* a fungus that causes wilts and death of many shade trees and ornamental shrubs, can be carried in infested mulch and kill susceptible plants in the landscape. *Rhizoctonia solani,* another plant pathogen that causes damping-off of many types of seedling plants, is actually stimulated by fresh mulches. This pathogen utilizes the cellulose in wood as a source of food.

Short-term composting of mulches in windrows under high-temperature conditions (130-160 degrees F) kills these plant pathogens. Six weeks of composting is sufficient to kill most plant pathogens and avoid their dissemination in mulches or composts.

Mycorrhizae, which are fungi that can form beneficial associations with roots, also are affected by mulches and composts. A shallow layer of wood chips (1-2") or compost improves tree establishment because mycorrhizae are stimulated by the slow release of organic sources of nitrogen and carbon in organic matter. However, a deep layer (4-6") of the same freshly chipped wood has been shown to inhibit the development of mycorrhizae during reforestation. Negative effects on mycorrhizae must be avoided in the landscape because they are very important in the maintenance of healthy plants.

Compost and mulch producers, landscapers, and homeowners can take measures to minimize fungal problems in the landscape. The type of mulch used, fresh versus composted mulch, the moisture content of the mulch before and during its utilization, the temperature and pH of the mulch before and during utilization, and the depth to which it is applied all play a role. Each factor is discussed here.

MULCH TYPE AND FRESH VERSUS COMPOSTED MULCH

Wood products from some trees are more resistant to decay than others and, therefore, cause fewer problems. Bark chips (nuggets) from large mature pine or other softwood trees such as cypress trees contain mostly lignin (dark material in bark), wax and protected cellulose that resist decay. On the other hand, wood wastes from these same tree species, but ground as young trees, rot quite readily because the cellulose in such bark and wood products is not yet protected from decomposition by lignin waxes or tannins.

Hardwood tree bark (oak, maple, etc.), even from large trees, contains a large concentration of cellulose that is not protected from rotting. Therefore, hardwood bark mulches, like ground wood from almost all tree species, rot readily and cause most of the nuisance mold problems in the landscape. The finer the product is ground, the more severe the problem can be! These materials are low in nitrogen content.

The fine particles (less than 3/4" diameter) in such mulches cause nitrogen immobilization in soil. The microflora that decomposes the wood particles takes up the nitrogen required for growth of plants. The result is that the plant becomes starved for nitrogen. Some mulch producers screen all particles smaller than 3/8" out of high-wood-content or hardwood-bark mulches, which avoids most of the nitrogen immobilization problem.

The best way to avoid all these problems and bring about beneficial effects by mulching is to add nitrogen to woody and hardwood bark products followed by composting to lower the carbon to nitrogen ratio. Blending of grass clippings with wood wastes before composting is one way to achieve this. Addition of poultry manure or urea to supply 1.2 lbs. available nitrogen per cubic yard of material satisfies the nitrogen need also. Some landscapers add 10-15 per cent by volume composted sewage sludge to hardwood bark or wood wastes, and this makes an ideal product that has performed very well in landscapes.

These amended products should be composted at least six weeks. This process kills plant pathogens, eggs of insect pests, and produces a nitrified product that releases plant nutrients rather than ties up nitrogen. As mentioned above, the microorganisms that have colonized these products reduce the potential for growth of nuisance fungi and provide control of many plant diseases.

TEMPERATURE, MOISTURE CONTENT, AND PH

Landscapers often apply quality mulch products from high temperature piles (140-160 degrees F) directly into the landscape. The temperature of the mulch is high because of heat produced by growth of microorganisms, known as thermophiles, in storage piles during the composting process. These microorganisms die soon after the mulch cools to 50-80oF after it has been placed around homes.

Because they require high temperatures to survive, they cannot grow and compete with soil microorganisms at the low temperature of mulches in the landscape. The sudden temperature drop that often occurs after mulch is applied creates what is known as a "biological vacuum." It also can occur during bagging of products at producers of mulches and particularly during dry seasons.

Mesophiles (low temperature soil microorganisms) rapidly colonize such mulches. If the mulch is dry, or dries out to a moisture content below 34 per cent during the first day after it is applied (mulches are dusty below this moisture content), fungi become the primary colonizers. This sets the stage for problems later, and the problem becomes most severe in mulches that are applied too deep in the landscape.

After prolonged heavy rains, the dry material colonized by fungi eventually becomes wet. Dry products stored in bags may also become moist when water produced as a result of microbial activity accumulates along the inner surface in bags. Bacteria then rapidly colonize the fungal white mass to induce the formation of fruiting structures by the fungi. The nuisance toad stools and other fruiting structures appear a few days later. Mold problems occur also when dry products are bagged or applied to dry soils.

Dry composts removed from high-temperature piles occasionally cause mushroom problems in bags and also in soils at nurseries. These moldy products inhibit plant growth in field soils as well as in potting mixes. They also cause wettability problems if dry conditions persist in the soil for a few weeks to give fungi a chance to become the dominant colonizers. Plants do not grow well in such moldy soils.

These problems can be reduced by soaking the high-temperature products with water as they are applied in the landscape or bagged. The high-moisture organic matter then becomes rapidly colonized by bacteria during the first few days.

These bacteria compete with fungi to reduce the potential for the development of major mold problems. This strategy has been successfully applied over the past decade to hardwood as well as softwood composts and mulches. It has controlled nuisance problems caused by many fungi in various parts of the United States and abroad.

The pH or acidity of the mulch is another important factor. Sour mulches that give off acrid odors may range in pH from as low as 2.5 up to 4.8. Highly acidic mulches are toxic to most plants and promote the growth of fungi. Bacteria that inhibit fungal growth cannot colonize mulches when the pH is lower than 5.2. The low pH and fungal problems are avoided if the raw material is nitrified and composted as described earlier.

In summary, water applied at the right time during composting, storage, and mulching can solve most of the fungal nuisance problems. It is best to maintain a water content higher than 40 per cent on a total weight basis. Again this allows bacteria as well as fungi to colonize the organic matter, and it sets up competition for nuisance molds. The moisture content of most organic products actually can be raised above 50 per cent and not present excessive weight problems during transport.

WHAT TO DO ONCE THE PROBLEMS OCCUR

Sometimes very little can be done to control nuisance fungi other than to spade the mulch into the surface soil layer followed by soaking with water. Another option is to remove the mulch, place it in a heap after thorough wetting to allow for self-heating to occur (110-140 degrees F). This will kill nuisance fungi. If fresh dry mulch is placed on top of mulch colonized by nuisance fungi, the problems may occur again the following year or even earlier.

The best control strategy for homeowners and landscapers is to purchase composted products low in wood content. Fresh, finely ground woody products should be avoided for many reasons unless composted first. Coarse fresh woody products are much less likely to cause problems unless applied too deep. It is important to soak all mulches immediately after they have been applied.

Generally, mulches should not be applied to a depth greater than two inches. Mulches and composts applied in this manner provide many types of beneficial effects rather than nuisance problems, or worse, plant diseases. Sour mulches should be avoided altogether.

MYCORRHIZAS IN THE URBAN LANDSCAPE

Mycorrhizas are anatomically intimate associations between fine (feeder) roots of plants and some special soilborne fungi. For this reason they are called symbioses. The plant, being capable of producing its own food through photosynthesis, can potentially survive without the fungus. The latter, however, is often completely dependent on the live plant for sustenance; thus, it is a parasite of the plant.

The association usually (but not always) results in a nutritional benefit to both the host plant and the fungus: the plant receives extra soil mineral nutrients, such as phosphorus and nitrogen, and extra water absorbed by the fungus growing out into the soil; the fungus obtains sugars and other vital substances from the plant. This kind of two-way beneficial interaction is called a "mutualistic association."

In nature, the vast majority of plants are affected by this kind of association with fungi, to the point that some scientists claim that "the study of plants without their mycorrhizas is the study of artefacts. The majority of plants, strictly speaking, do not have roots, they have mycorrhizas.". In fact, it is now widely accepted that in the course of evolution, colonization of land by plants occurred thanks to these plant-fungus associations.

Mycorrhizas are important because plants grown in the absence of their fungal associates generally do not perform as well as those that are mycorrhizal. Sometimes, such as in the case of orchids, the seeds do not even germinate in the absence of the appropriate fungus.

Mycorrhizas are also known to protect the plants from various root diseases. However, lack of mycorrhizal roots on normally mycorrhizal plants is a rather rare occurrence, as most natural soils contain a full complement of mycorrhizal fungal propagules, that is, spores or other structures used by the fungi to infect new plants.

Mycorrhizas Come in Different Types

There are seven different kinds of mycorrhizas. The most important ones from an urban landscape perspective are the ectomycorrhizas, the endomycorrhizas, and the ericoid mycorrhizas. The main and most visible characteristic of ectomycorrhizas is the fungus growing on the surface (hence "ecto," which means "outside") of the feeder roots. This fungal growth is called a sheath or mantle.

The fungus grows from the mantle into the cortical layers of the root, but remains between the plant cells and does not penetrate them, forming a so-called Hartig net. This is where the nutrient exchanges between the plant and the fungus occur. On the other hand, endomycorrhizas are characterized by the absence of a mantle and by penetration of the fungus into the cortical cells.

Once inside the cells, the fungus produces characteristic structures, called "arbuscules" (= tiny trees) and "vesicles." The arbuscules are the structures where the nutrient exchanges between the plant and the fungus occur. Hence, endomycorrhizas are also called "vesicular arbuscular mycorrhizas" (VAM) or just "arbuscular mycorrhizas" (AM). Ericoid mycorrhizas are somewhat similar to VAM, in that the fungus invades the host cells, but affect only plants in the family Ericaceae.

Different plant groups tend to form specific types of associations. For example, Ericaceae tend to form ericoid mycorrhizas, P inaceae almost always form ectomycorrhizas, and hardwoods can form both ecto- and endomycorrhizas, sometimes on the same plant. Table shows a partial list of forest and ornamental trees and their specific type of association.

One last characteristic to bear in mind, particularly when considering the following paragraph, is that the fungi involved in the two main types of mycorrhizas, that is endo- and ectomycorrhizas, feature remarkably different levels of host species specificity.

While most individual ectomycorrhizal fungal species tend to form associations with specific host plant species or a restricted number of them, individual endomycorrhizal fungal species are "generalists" and can associate with hundreds of different host plant species. This characteristic is reflected in the number of recognized mycorrhizal fungal species: thousands of ectomycorrhizal vs. about 200 endomycorrhizal fungal species.

ARTIFICIAL INOCULATION

This question is becoming more and more common as more companies have recently begun commercializing mycorrhizal inoculum. Artificial mycorrhizal inoculation may benefit plants only when there is no natural and appropriate mycorrhizal inoculum in the soil, or the inoculum level is low, or species present are less efficient at aiding the plant host than those being introduced. This situation is most commonly found in highly impoverished soils that do not support natural fungal growth, for example in the case of mine spoils.

In principle, highly compacted, organic nutrient-poor soils in urban environments (often the result of mineral subsoil horizons having been turned into the top layers by construction activities) may also be prime candidates for artificial mycorrhizal inoculation. However, success of the treatment can never be guaranteed for two main reasons:

- The urban soil may already contain a complement of mycorrhizal fungi, however depleted, that outcompete the newly introduced species. Competition among mycorrhizal fungi is well documented and a common occurrence in natural soils, for example in reforestation operations. It is particularly important for ectomycorrhizal fungi;
- The soil organic matter content may not be adequate. The presence of organic matter can greatly enhance the development of mycorrhizas.

Due to the specificity properties of the different mycorrhizal associations described above, it is likely that mycorrhizal inoculation will be more successful, in terms of actual establishment of the inoculated species, with endomycorrhizal fungal species, even though the main ectomycorrhizal inoculum being commercialized is a "generalist.

Among specialists." At present, there is very limited, unbiased scientific evidence that artificial mycorrhizal inoculum in the landscape makes plant establishment more successful or that the inoculated plants grow better over time. In fact, the available evidence is rather inconsistent. In part this is due to the fact that commercial preparations may not be very viable or not viable at all by the time they reach the consumer, or become not viable once they are in the consumer's hands.

But perhaps the biggest factor is that soil conditions in the landscape may be too variable to be able to make generalizations about the usefulness of artificial inoculation with the few selected species available commercially.

Furthermore, any short-term positive effects observed on plant growth following mycorrhizal inoculation may simply be the result of the fertilizer components that are often present in the commercial preparations. The bottom line is that artificial inoculation may have either a positive or neutral effect on plant growth and health. In all cases, however, it is very unlikely that artificial inoculation is detrimental to the plant.

In other words, the treatment may or may not work, but would not be injurious. Current research, conducted in the author's and other researchers' laboratories, suggests that organic soil amendments, particularly the addition of composted mulches, greatly enhances the mycorrhizal status of landscape plants, even without the addition of artificial inoculum. Improvement of soil conditions may therefore be the most important factor in mycorrhizal development in the urban landscape.

OAK WILT

Oak wilt is a serious and often deadly vascular disease of oaks. The fungal pathogen, Ceratocystis fagacearum, is believed to be native to the United States and is distributed throughout the Midwest and Texas.

What the Pathogen Does

The fungus grows into and throughout the water conductive tissues (that is, the sapwood) of the host. The fungus plugs the vessels with its own body (mycelium) and spores, but it also causes a defensive reaction by the tree to stop the fungal spread by actively plugging its own vessels. These processes interfere with water uptake and cause a wilting syndrome which often results in death of the tree.

Susceptible Oaks

All oaks are susceptible. Those in the red-black oak group (black, blackjack, pin, northern and southern red, scarlet, shingle and shumard oak) are extremely susceptible and can die within a few weeks of infection. Oaks in the white group (bur, chinquapin, post, swamp white, and white oak) are more tolerant of the disease and may survive infection for one or more years while displaying decline symptoms.

Diagnostic Symptoms

Symptoms are typical of wilts. Leaves usually begin withering in the upper canopy, producing "flags," that is, whole branches or crown portions turning red-brown. Leaves of red oaks typically show yellowing and browning of the leaf margins. White oak leaves usually show rather non-descript symptoms. Conversely, live oaks in the southern United States produce characteristic dead areas along the leaf veins.

These dead areas generally expand until the whole leaf becomes brown. Eventually the leaves fall from the tree. If infections occur in late spring, trees usually begin wilting in mid-summer to late summer, when the plants often are subjected to water deficit due to increased transpiration demand and decreased rainfall.

A specific and sufficient diagnostic character is the appearance on dead and dying red oaks (but not white oaks) of spore-bearing fungal mats under the desiccating bark. These fungal mats crack the bark open with pressure pads to facilitate dissemination of the pathogen. Sapwood streaking is also a good, but insufficient, diagnostic character.

Disease Cycle and Conditions Favouring Disease

In order to properly manage oak wilt it is essential to understand its cycle. The pathogen spreads from diseased to healthy trees in two ways: overland and underground. Overland spread is mediated mainly by sap feeding (a.k.a. picnic) beetles (Coleoptera: Nitidulidae). However, there is some evidence that oak bark beetles (Coleoptera: Scolytidae) may also be involved. Nitidulids are attracted by chemicals emanating from the fungal mats described above.

Once on the mats, the beetles pick up fungal spores and can carry them, sometimes over distances of a few miles, to freshly wounded healthy trees (attracted by the smell of fresh sap). This results in new infections, thus closing the overland cycle. While insect spread is an important medium to long range dispersal mechanism for this fungus, it is estimated that 90 per cent of new infections occur between neighbouring trees through root grafts. In this case, the fungus grows down the trunk, into the roots of diseased trees, and then into healthy trees via the common root system. Once in the new tree the pathogen grows throughout the vascular system and spreads to other trees via the root system or the beetles. In this way, spread through root systems often results in disease centres that expand outward from the initially infected tree.

From the above, it follows that conditions favouring disease include the availability of susceptible oak species, trees growing close to each other, and the availability of fresh wounds for beetle-mediated infection. Pruning wounds are obvious culprits, but any fresh wound will function as potential infection gateway. The word fresh is emphasized because it is believed that wounds are attractive to Nitidulid beetles only for up to three days. As with many plant diseases, other stresses (for example drought) can predispose trees to faster symptom development, and thus worsen the syndrome.

Control and Management of the Disease

The best control for oak wilt is through preventative measures that interrupt the disease cycle.

Prevention of Overland Spread

Overland spread can be hindered or interrupted by ensuring that trees are never wounded between April 15 and July 1. This is when most Nitidulid beetles fly to locate fresh sap and/or fungal mats. A more stringent approach is to avoid wounding the trees throughout the growing season, since additional summer flights of the beetles are possible.

If pruning is absolutely necessary during the growing season, it is imperative to dress the wounds. This can be done with latex paint. Although this will slow wound healing, it will also deter beetles from landing on the wounds.

Prevention of Underground Spread

Given the higher significance of underground spread,

control of direct tree-to-tree transmission is much more important. Here, interruption of the disease cycle is accomplished by physically severing actual or potential root contacts between diseased and healthy trees. This is done by trenching or cutting through the soil with a trencher or vibratory plow. The latter is the preferred tool. Given the depth of oak root systems, it is advisable to use a 5 ft blade.

Trenching must always be done before the diseased or dead trees are cut for removal, to avoid sudden water tension imbalances that might "suck" fungal material from the infected trees into the healthy trees through the common root system. Trenching should be conducted by advice of specialists. This is due to the importance of locating the trenches appropriately between diseased and healthy trees. When possible, a double trench defining a buffer band of apparently healthy trees between diseased and uninfected trees should be used.

On residential or commercial properties, always determine the location of buried utility lines which may affect the ability to completely sever the graft. Furthermore, walkways, paths, and roads must also be considered appropriately, as tree roots commonly grow under them. Due to all of these potential obstacles to proper trenching, it is advisable to undertake such operations under the supervision of tree care professionals with expertise in the management of oak wilt.

There is currently no evidence that the blade will spread the pathogen. However, it is good precautionary practice to spray the blade to runoff, between trenches and between plots, with an antiseptic such as Lysol or a 20 per cent bleach solution.

Disposal of Dead and Dying Trees

Once the trenches are in place, diseased and dead trees should be removed as soon as possible by cutting them down to leave a 2-4 inch high stump. Because diseased trees with bark tightly attached may produce or harbor fungal mats, they should be disposed of promptly. Once the bark becomes loose or sloughs off, no mats can be produced and movement of the infected wood out of the diseased area is no longer a concern in an urban context. (However, strict restrictions apply for movement of diseased wood out of state and internationally.) Thus, either the trees are debarked mechanically, or the timber can be sold to a sawmill for cutting or chipping.

Although no studies are known on the transmissibility of C. fagacearum via wood chips, the pathogen does not survive when exposed to desiccation and is very temperature sensitive. Composting the chips would further reduce or eliminate the pathogen. Thus, it is highly unlikely that wood chips will spread the disease. However, good precautionary practice suggests to avoid using infested chip mulch around healthy oaks.

If the wood cannot be disposed of as described above, it can be cut and split for firewood. Because this process does not involve debarking, firewood can still potentially harbor fungal mats and thus attract Nitidulids during the summer in which the trees died. The wood must be arranged in stacks and covered with 4 mil plastic tarp through the winter (if the wood is used then) or the end of the next season (Oct. 1 of the year following the death of the trees).

By producing a greenhouse effect, tarping will kill the temperature-sensitive pathogen and prevent the beetles from accessing potential fungal mats. Tarping should be done with transparent plastic to produce the desired greenhouse effect. However, black plastic will also work, by concentrating the sun's heat.

In both cases, the best results are achieved by placing the tarped pile in an un-shaded, possibly sunny area. When covering the pile, the tarp should be sealed to the ground to prevent beetles from accessing the pile. For this reason, all punctures in the tarp should be mended with duct tape. At the end of the second season the wood can be safely uncovered and disposed of as preferred, since it no longer constitutes a threat.

Chemical Treatments

Chemical treatments are usually not warranted, due to the high cost of intervention. However, application of systemic fungicides is an option when highly valuable trees are threatened by infected neighbouring trees, or whenever a high risk of infection exists.

High value may be attributed to individual trees or groups of trees in communities, or it may apply to individual trees in a homeowner's yard. Systemic fungicides have been demonstrated to be effective, particularly when applied as a preventative treatment. The only scientifically tested systemic fungicide showing any efficacy and labeled for use against oak wilt is propiconazole, available under the trade name Alamo. If a decision is made to apply propiconazole, it must be done strictly according to label and should be carried out by experienced, professional tree service personnel. The product is injected directly into the sapwood on root flares just under the soil line.

This is the best guarantee that the fungicide will be translocated throughout a tree, thus affording the maximum possible protection. In consideration of the disease cycle, the best time of the year to inject trees is early spring. However, application should occur as soon as the risk to a tree is realised, even if it is later in the growing season.

Depending on the tree size and value, treatments should be applied every 12-36 months, with annual assessments. This treatment has virtually no hope of succeeding in infected red oaks, even in early infection stages.

Chemical treatment has a higher chance of success, but only as a palliative measure, with the more tolerant white oaks. In this group, application of the fungicide to trees in early infection stages can result in delay of symptoms and eventual death. It will not, however, rid an infected tree of the pathogen.

ANTHRACNOSE FRUIT ROT OF PEPPER

Several species of plant pathogenic fungi in the genus *Colletotrichum* cause anthracnose in peppers and many other vegetables and fruits. Until the late 1990s, anthracnose of peppers and tomatoes was only associated with ripe or ripening fruit.

Since that time, a more aggressive form of the disease has become established. This form attacks peppers at any stage of fruit development and may threaten the profitability of pepper crops in areas where it becomes established. This disease can also affect tomatoes, strawberries, and possibly other fruit and vegetable crops.

Symptoms

Circular or angular sunken lesions develop on immature fruit of any size. Often multiple lesions form on individual fruit. When disease is severe, lesions may coalesce. Often pink to orange masses of fungal spores form in concentric rings on the surface of the lesions.

In older lesions, black structures called acervuli may be observed. With a hand lens, these look like small black dots; under a microscope they look like tufts of tiny black hairs. The pathogen forms spores quickly and profusely and can spread rapidly throughout a pepper crop, resulting in up to 100 per cent yield loss. Lesions may also appear on stems and leaves as irregularly shaped brown spots with dark brown edges.

Causal Organism and Disease Cycle

This form of pepper anthracnose is caused by the fungus *Colletotrichum acutatum*. The pathogen survives on plant debris from infected crops and on other susceptible plant species. The fungus is not soil-borne for long periods in the absence of infested plant debris.

The fungus may also be introduced into a crop on infested seed. During warm and wet periods, spores are splashed by rain or irrigation water from diseased to healthy fruit. Diseased fruit act as a source of inoculum, allowing the disease to spread from plant to plant within the field.

- Plant only seed from disease-free plants or seed treated to reduce any fungal populations. Seed can be disinfested with a 30-minute soak at 52°C. for complete seed-treatment instructions.
- Peppers should be rotated out of infested fields for

at least three years. Fields should be planted with crops other than tomatoes, eggplants, or other solanaceous crops or strawberries, which are also hosts of *C. acutatum*.

- Apply overhead irrigation during the early part of the day so that plants can dry before sundown.
- In areas where market constraints and other diseases do not limit the choice of cultivar, cultivars demonstrating a moderate level of resistance should be chosen.
- Apply protective fungicides at flowering when environmental conditions are favourable for disease development. Rates and timing of all further applications should be done according to product label instructions.

Index

D

E

F

G

H

I

J

K

L

M